CONTINUUM MECHANICS

A. J. M. Spencer

Emeritus Professor of Theoretical Mechanics
University of Nottingham

DOVER PUBLICATIONS, INC.
Mineola, New York

Copyright

Bibliographical Note

This Dover edition, first published in 2004, is an unabridged republication of the edition originally published by the Longman Group UK Limited, Essex, England, in 1980.

Library of Congress Cataloging-in-Publication Data

Spencer, A. J. M. (Anthony James Merrill), 1929–
Continuum mechanics / A.J.M. Spencer.
p. cm.
Originally published: London ; New York : Longman, 1980. (Longman mathematical texts)
Includes bibliographical references and index.
ISBN-13: 978-0-486-43594-7 (pbk.)
ISBN-10: 0-486-43594-6 (pbk.)
1. Continuum mechanics. I. Title.

QA808.2.S63 2004
531—dc22

2003070116

www.doverpublications.com

Contents

Preface

The aim of this book is to provide an introduction to the theory of continuum mechanics in a form which is suitable for undergraduate students. It is based on lectures which I have given in the University of Nottingham during the last fourteen years. I have tried to restrict the mathematical background required to that which is normally familiar to a second-year mathematics undergraduate or a mathematically minded engineering graduate, even though some of the theory can be developed more concisely and elegantly by using more sophisticated mathematics than I have employed. The material covered comprises introductory chapters on matrix algebra and on vectors and cartesian tensors, the analysis of deformation and stress, the mathematical statements of the laws of conservation of mass, momentum and energy, and the formulation of the mechanical constitutive equations for various classes of fluids and solids. Cartesian coordinates and cartesian tensors are used throughout, except that in the last chapter we show how the theory can be expressed in terms of cylindrical polar and spherical polar coordinates. I have not pursued the various branches of the mechanics of solids and fluids, such as elasticity, Newtonian fluid mechanics, viscoelasticity and plasticity, beyond the point of formulating their constitutive equations. To do so in any meaningful way would have required a much longer book, and these subjects are fully dealt with in larger and more specialized texts.

I am, of course, greatly indebted to many teachers, colleagues and students who have contributed to my education in continuum mechanics. They are too numerous to mention individually; rather than giving a selective list I ask them to accept a collective acknowledgement. Similarly, I have felt that in an introductory book of this kind it would be inappropriate to give references to original work, but it is obvious that I have made indirect use of many sources and I am glad to acknowledge the contribution of all the authors whose work has influenced me.

Many of the problems are taken from examination papers set in the Department of Theoretical Mechanics in the University of Nottingham, and I acknowledge the University's permission to make use of these.

Finally, I thank Margaret for the typing.

A. J. M. SPENCER
Nottingham, 1979

1 Introduction

1.1 Continuum mechanics

Modern physical theories tell us that on the microscopic scale matter is discontinuous; it consists of molecules, atoms and even smaller particles. However, we usually have to deal with pieces of matter which are very large compared with these particles; this is true in everyday life, in nearly all engineering applications of mechanics, and in many applications in physics. In such cases we are not concerned with the motion of individual atoms and molecules, but only with their behaviour in some average sense. In principle, if we knew enough about the behaviour of matter on the microscopic scale it would be possible to calculate the way in which material behaves on the macroscopic scale by applying appropriate statistical procedures. In practice, such calculations are extremely difficult; only the simplest systems can be studied in this way, and even in these simple cases many approximations have to be made in order to obtain results. Consequently, our knowledge of the mechanical behaviour of materials is almost entirely based on observations and experimental tests of their behaviour on a relatively large scale.

Continuum mechanics is concerned with the mechanical behaviour of solids and fluids on the macroscopic scale. It ignores the discrete nature of matter, and treats material as uniformly distributed throughout regions of space. It is then possible to define quantities such as density, displacement, velocity, and so on, as continuous (or at least piecewise continuous) functions of position. This procedure is found to be satisfactory provided that we deal with bodies whose dimensions are large compared with the characteristic lengths (for example, interatomic spacings in a crystal, or mean free paths in a gas) on the microscopic scale. The microscopic scale need not be of atomic dimensions; we can, for example, apply continuum mechanics to a granular material such as sand, provided that the dimensions of the region considered are large compared with those of an individual grain. In continuum mechanics it is assumed that we can associate a particle of matter with each and every point of the region of space occupied by a body, and ascribe field quantities such as density, velocity, and so on, to these particles. The justification for this procedure is to some extent based on statistical mechanical theories of gases, liquids and solids, but rests mainly on its success in describing and predicting the mechanical behaviour of material in bulk.

Mechanics is the science which deals with the interaction between force and motion. Consequently, the variables which occur in continuum mechanics are, on the one hand, variables related to forces (usually force per unit area or per unit volume, rather than force itself) and, on the other hand, kinematic variables such as displacement, velocity and acceleration. In rigid-body mechanics, the shape of a body does not change, and so the particles which make up a rigid body may only move relatively to one another in a very restricted way. A rigid body is a continuum, but it is a very special, idealized and untypical one. Continuum mechanics is more concerned with deformable bodies, which are capable of changing their shape. For such bodies the relative motion of the particles is important, and this introduces as significant kinematic variables the spatial derivatives of displacement, velocity, and so on.

The equations of continuum mechanics are of two main kinds. Firstly, there are equations which apply equally to all materials. They describe universal physical laws, such as conservation of mass and energy. Secondly, there are equations which describe the mechanical behaviour of particular materials; these are known as constitutive equations.

The problems of continuum mechanics are also of two main kinds. The first is the formulation of constitutive equations which are adequate to describe the mechanical behaviour of various particular materials or classes of materials. This formulation is essentially a matter for experimental determination, but a theoretical framework is neeeded in order to devise suitable experiments and to interpret experimental results. The second problem is to solve the constitutive equations, in conjunction with the general equations of continuum mechanics, and subject to appropriate boundary conditions, to confirm the validity of the constitutive equations and to predict and describe the behaviour of materials in situations which are of engineering, physical or mathematical interest. At this problem-solving stage the different branches of continuum mechanics diverge, and we leave this aspect of the subject to more comprehensive and more specialized texts.

2 Introductory matrix algebra

2.1 Matrices

In this chapter we summarize some useful results from matrix algebra. It is assumed that the reader is familiar with the elementary operations of matrix addition, multiplication, inversion and transposition. Most of the other properties of matrices which we will present are also elementary, and some of them are quoted without proof. The omitted proofs will be found in standard texts on matrix algebra.

An $m \times n$ matrix $\mathbf{A}$ is an ordered rectangular array of mn elements. We denote

$$\mathbf{A} = (A_{ij}) = \begin{pmatrix} A_{11} & A_{12} & \dots & A_{1n} \\ A_{21} & A_{22} & \dots & A_{2n} \\ \cdot & \cdot & & \cdot \\ \cdot & \cdot & & \cdot \\ \cdot & \cdot & & \cdot \\ A_{m1} & A_{m2} & \dots & A_{mn} \end{pmatrix} \tag{2.1}$$

so that A_{ij} is the element in the ith row and the jth column of the matrix $\mathbf{A}$. The *index* i takes values $1, 2, \dots, m$, and the index j takes values $1, 2, \dots, n$. In continuum mechanics the matrices which occur are usually either 3×3 square matrices, 3×1 column matrices or 1×3 row matrices. We shall usually denote 3×3 square matrices by bold-face roman capital letters ($\mathbf{A}$, $\mathbf{B}$, $\mathbf{C}$, etc.) and 3×1 column matrices by bold-face roman lower-case letters ($\mathbf{a}$, $\mathbf{b}$, $\mathbf{c}$, etc.). A 1×3 row matrix will be treated as the transpose of a 3×1 column matrix ($\mathbf{a}^{\mathrm{T}}$, $\mathbf{b}^{\mathrm{T}}$, $\mathbf{c}^{\mathrm{T}}$, etc.). Unless otherwise stated, indices will take the values 1, 2 and 3, although most of the results to be given remain true for arbitrary ranges of the indices.

A square matrix $\mathbf{A}$ is *symmetric* if

$$\mathbf{A} = \mathbf{A}^{\mathrm{T}}, \qquad A_{ij} = A_{ji} \tag{2.2}$$

and *anti-symmetric* if

$$\mathbf{A}=-\mathbf{A}^{\mathrm{T}} \qquad A_{ij}=-A_{ji} \tag{2.3}$$

where $\mathbf{A}^{\mathrm{T}}$ denotes the transpose of $\mathbf{A}$.

The 3×3 *unit matrix* is denoted by $\mathbf{I}$, and its elements by δ_{ij}. Thus

$$\mathbf{I}=(\delta_{ij}) \qquad (i,j=1,2,3) \tag{2.4}$$

where

$$\delta_{11}=\delta_{22}=\delta_{33}=1, \qquad \delta_{23}=\delta_{31}=\delta_{12}=\delta_{32}=\delta_{13}=\delta_{21}=0 \tag{2.5}$$

Clearly $\delta_{ij}=\delta_{ji}$. The symbol δ_{ij} is known as the *Kronecker delta.* An important property of δ_{ij} is the substitution rule:

$$\sum_{j=1}^{3} \delta_{ij}A_{jk}=A_{ik}, \qquad \sum_{j=1}^{3} \delta_{ij}A_{kj}=A_{ki} \tag{2.6}$$

The *trace* of a square matrix $\mathbf{A}$ is denoted by $\operatorname{tr}\mathbf{A}$, and is the sum of the elements on the leading diagonal of $\mathbf{A}$. Thus, for a 3×3 matrix $\mathbf{A}$,

$$\operatorname{tr}\mathbf{A}=A_{11}+A_{22}+A_{33}=\sum_{i=1}^{3} A_{ii} \tag{2.7}$$

In particular,

$$\operatorname{tr}\mathbf{I}=\sum_{i=1}^{3} \delta_{ii}=3 \tag{2.8}$$

With a square matrix $\mathbf{A}$ there is associated its *determinant*, $\det\mathbf{A}$. We assume familiarity with the elementary properties of determinants. The determinant of a 3×3 matrix $\mathbf{A}$ can be expressed as

$$\det\mathbf{A}=\tfrac{1}{6}\sum_{i=1}^{3}\sum_{j=1}^{3}\sum_{k=1}^{3}\sum_{r=1}^{3}\sum_{s=1}^{3}\sum_{t=1}^{3} e_{ijk}e_{rst}A_{ir}A_{js}A_{kt} \tag{2.9}$$

where the *alternating symbol* e_{ijk} is defined as:

(a) $e_{ijk}=1$ if (i,j,k) is an even permutation of $(1,2,3)$ (i.e. $e_{123}=e_{231}=e_{312}=1$);
(b) $e_{ijk}=-1$ if (i,j,k) is an odd permutation of $(1,2,3)$ (i.e. $e_{321}=e_{132}=e_{213}=-1$);
(c) $e_{ijk}=0$ if any two of i, j, k are equal (e.g. $e_{112}=0$, $e_{333}=0$).

It follows from this definition that e_{ijk} has the symmetry properties

$$e_{ijk} = e_{jki} = e_{kij} = -e_{kji} = -e_{ikj} = -e_{jik} \tag{2.10}$$

The condition $\det \mathbf{A} \neq 0$ is a necessary and sufficient condition for the existence of the inverse $\mathbf{A}^{-1}$ of $\mathbf{A}$.

A square matrix $\mathbf{Q}$ is *orthogonal* if it has the property

$$\mathbf{Q}^{-1} = \mathbf{Q}^{\mathrm{T}} \tag{2.11}$$

It follows that if $\mathbf{Q}$ is orthogonal, then

$$\mathbf{Q}\mathbf{Q}^{\mathrm{T}} = \mathbf{I}, \qquad \mathbf{Q}^{\mathrm{T}}\mathbf{Q} = \mathbf{I} \tag{2.12}$$

and

$$\det \mathbf{Q} = \pm 1 \tag{2.13}$$

Our main concern will be with proper orthogonal matrices, for which

$$\det \mathbf{Q} = 1$$

If $\mathbf{Q}_1$ and $\mathbf{Q}_2$ are two orthogonal matrices, then their product $\mathbf{Q}_1\mathbf{Q}_2$ is also an orthogonal matrix.

2.2 The summation convention

A very useful notational device in the manipulation of matrix, vector and tensor expressions is the summation convention. According to this, if the same index occurs twice in any expression, summation over the values 1, 2 and 3 of that index is automatically assumed, and the summation sign is omitted. Thus, for example, in (2.7) we may omit the summation sign and write

$$\operatorname{tr} \mathbf{A} = A_{ii}$$

Similarly, the relations (2.6) are written as

$$\delta_{ij} A_{jk} = A_{ik}, \qquad \delta_{ij} A_{kj} = A_{ki}$$

and from (2.8),

$$\delta_{ii} = 3$$

Using this convention, (2.9) becomes

$$\det \mathbf{A} = \tfrac{1}{6} e_{ijk} e_{rst} A_{ir} A_{js} A_{kt} \tag{2.14}$$

The conciseness introduced by the use of this notation is illustrated by the observation that, in full, the right-hand side of (2.14) contains $3^6 = 729$ terms, although because of the properties of e_{ijk} only six of these are distinct and non-zero.

Some other examples of the use of summation convention are the following:

(a) If $\mathbf{A} = (A_{ij})$, $\mathbf{B} = (B_{ij})$, then the element in the ith row and jth column of the product $\mathbf{AB}$ is $\sum_{k=1}^{3} A_{ik}B_{kj}$, which is written as $A_{ik}B_{kj}$.

(b) Suppose that in (a) above, $\mathbf{B} = \mathbf{A}^{\mathrm{T}}$. Then $B_{ij} = A_{ji}$, and so the element in the ith row and jth column of $\mathbf{AA}^{\mathrm{T}}$ is $A_{ik}A_{jk}$. In particular, if $\mathbf{A}$ is an orthogonal matrix $\mathbf{Q} = (Q_{ij})$ we have from (2.12)

$$Q_{ik}Q_{jk} = \delta_{ij}, \qquad Q_{ki}Q_{kj} = \delta_{ij} \tag{2.15}$$

(c) A linear relation between two column matrices $\mathbf{x}$ and $\mathbf{y}$ has the form

$$\mathbf{x} = \mathbf{Ay} \tag{2.16}$$

which may be written as

$$x_i = A_{ij}y_j \tag{2.17}$$

If $\mathbf{A}$ is non-singular, then from (2.16), $\mathbf{y} = \mathbf{A}^{-1}\mathbf{x}$. In particular, if $\mathbf{A}$ is an orthogonal matrix $\mathbf{Q}$, then

$$\mathbf{x} = \mathbf{Qy}, \qquad x_i = Q_{ij}y_j,$$
$$\mathbf{y} = \mathbf{Q}^{-1}\mathbf{x} = \mathbf{Q}^{\mathrm{T}}\mathbf{x}, \qquad y_i = Q_{ji}x_j.$$

(d) The trace of $\mathbf{AB}$ is obtained by setting $i = j$ in the last expression in (a) above; thus

$$\operatorname{tr} \mathbf{AB} = A_{ik}B_{ki} \tag{2.18}$$

By a direct extension of this argument

$$\operatorname{tr} \mathbf{ABC} = A_{ij}B_{jk}C_{ki},$$

and so on.

(e) If $\mathbf{a}$ and $\mathbf{b}$ are column matrices with

$$\mathbf{a}^{\mathrm{T}} = (a_1 \quad a_2 \quad a_3), \qquad \mathbf{b}^{\mathrm{T}} = (b_1 \quad b_2 \quad b_3),$$

then $\mathbf{a}^T\mathbf{b}$ is a 1×1 matrix whose single element is

$$a_1b_1+a_2b_2+a_3b_3=\sum_{i=1}^{3} a_ib_i=a_ib_i \tag{2.19}$$

(f) If $\mathbf{a}$ is as in (e) above, and $\mathbf{A}$ is a 3×3 matrix, then $\mathbf{Aa}$ is a 3×1 column matrix, and the element in its ith row is $\sum_{r=1}^{3} A_{ir}a_r$, which is written as $A_{ir}a_r$.

(g) Two useful relations between the Kronecker delta and the alternating symbol are

$$\begin{aligned} e_{ijp}e_{ijq} &= 2\delta_{pq} \\ e_{ijp}e_{rsp} &= \delta_{ir}\delta_{js}-\delta_{is}\delta_{jr} \end{aligned} \tag{2.20}$$

These can be verified directly by considering all possible combinations of values of i, j, p, q, r and s. Actually, (2.20) are consequences of a more general relation between δ_{ij} and e_{ijk}, which can also be proved directly, and is

$$\begin{vmatrix} \delta_{ir} & \delta_{is} & \delta_{it} \\ \delta_{jr} & \delta_{js} & \delta_{jt} \\ \delta_{kr} & \delta_{ks} & \delta_{kt} \end{vmatrix} = e_{ijk}e_{rst} \tag{2.21}$$

From (2.14) and (2.21) we can obtain the useful relation

$$e_{mpq}\det\mathbf{A}=e_{ijk}A_{im}A_{jp}A_{kq} \tag{2.22}$$

An index on which a summation is carried out is called a *dummy index*. A dummy index may be replaced by any other dummy index, for example $A_{ii}=A_{jj}$. However, it is important always to ensure that, when the summation convention is employed, no index appears more than twice in any expression, because the expression is then ambiguous.

In the remainder of this book it is to be assumed, unless the contrary is stated, that the summation convention is being employed. This applies, in subsequent chapters, to indices which label vector and tensor components as well as those which label matrix elements.

2.3 Eigenvalues and eigenvectors

In continuum mechanics, and in many other subjects, we frequently encounter homogeneous algebraic equations of the form

$$\mathbf{Ax}=\lambda\mathbf{x} \tag{2.23}$$

where $\mathbf{A}$ is a given square matrix, $\mathbf{x}$ an unknown column matrix and λ an unknown scalar. In the applications which appear in this book, $\mathbf{A}$ will be a 3×3 matrix. We therefore confine the discussion to the case in which $\mathbf{A}$ is a 3×3 matrix, although the generalization to $n\times n$ matrices is straightforward. Equation (2.23) can be written in the form

$$(\mathbf{A}-\lambda\mathbf{I})\mathbf{x}=\mathbf{0} \tag{2.24}$$

and the condition for (2.24) to have non-trivial solutions for $\mathbf{x}$ is

$$\det(\mathbf{A}-\lambda\mathbf{I})=0 \tag{2.25}$$

This is the *characteristic equation* for the matrix $\mathbf{A}$. When the determinant is expanded, (2.25) becomes a cubic equation for λ, with three roots λ_1, λ_2, λ_3 which are called the *eigenvalues* of $\mathbf{A}$. For the present we assume that λ_1, λ_2 and λ_3 are distinct. Then, for example, the equation

$$(\mathbf{A}-\lambda_1\mathbf{I})\mathbf{x}=0$$

has a non-trivial solution $\mathbf{x}^{(1)}$, which is indeterminate to within a scaler multiplier. The column matrix $\mathbf{x}^{(1)}$ is the *eigenvector* of $\mathbf{A}$ associated with the eigenvalue λ_1; eigenvectors $\mathbf{x}^{(2)}$ and $\mathbf{x}^{(3)}$ associated with the eigenvalues λ_2 and λ_3 are defined similarly.

Since λ_1, λ_2, λ_3 are the roots of (2.25), and the coefficient of λ^3 on the left of (2.25) is -1, we have

$$\det(\mathbf{A}-\lambda\mathbf{I})=(\lambda_1-\lambda)(\lambda_2-\lambda)(\lambda_3-\lambda) \tag{2.26}$$

This is an identity in λ, so it follows by setting $\lambda=0$ that

$$\det\mathbf{A}=\lambda_1\lambda_2\lambda_3 \tag{2.27}$$

Now suppose that $\mathbf{A}$ is a real symmetric matrix. There is no *a priori* reason to expect λ_1 and $\mathbf{x}^{(1)}$ to be real. Suppose they are complex, with complex conjugates $\bar{\lambda}_1$ and $\bar{\mathbf{x}}^{(1)}$. Then

$$\mathbf{A}\mathbf{x}^{(1)}=\lambda_1\mathbf{x}^{(1)} \tag{2.28}$$

Transposing (2.28) and taking its complex conjugate gives

$$\bar{\mathbf{x}}^{(1)\mathrm{T}}\mathbf{A}=\bar{\lambda}_1\bar{\mathbf{x}}^{(1)\mathrm{T}} \tag{2.29}$$

Now multiply (2.28) on the left by $\bar{\mathbf{x}}^{(1)\mathrm{T}}$ and (2.29) on the right by $\mathbf{x}^{(1)}$, and subtract. This gives

$$(\lambda_1-\bar{\lambda}_1)\bar{\mathbf{x}}^{(1)\mathrm{T}}\mathbf{x}^{(1)}=\mathbf{0} \tag{2.30}$$

Since $\mathbf{x}^{(1)}$ is a non-trivial solution of (2.24), $\bar{\mathbf{x}}^{(1)\mathrm{T}}\mathbf{x}^{(1)} \neq \mathbf{0}$, and so $\lambda_1 = \bar{\lambda}_1$. Hence *the eigenvalues of a real symmetric matrix are real.*

Also from (2.28),

$$\mathbf{x}^{(2)\mathrm{T}}\mathbf{A}\mathbf{x}^{(1)} = \lambda_1 \mathbf{x}^{(2)\mathrm{T}}\mathbf{x}^{(1)} \tag{2.31}$$

and similarly

$$\mathbf{x}^{(1)\mathrm{T}}\mathbf{A}\mathbf{x}^{(2)} = \lambda_2 \mathbf{x}^{(1)\mathrm{T}}\mathbf{x}^{(2)} \tag{2.32}$$

Now transpose (2.31) and subtract the resulting equation from (2.32). This gives

$$(\lambda_2 - \lambda_1)\mathbf{x}^{(1)\mathrm{T}}\mathbf{x}^{(2)} = \mathbf{0} \tag{2.33}$$

Hence the eigenvectors associated with two *distinct* eigenvalues λ_1 and λ_2 of a *symmetric* matrix $\mathbf{A}$ have the property $\mathbf{x}^{(1)\mathrm{T}}\mathbf{x}^{(2)} = \mathbf{0}$. Two column matrices with this property are said to be *orthogonal*. In general, if the eigenvalues are distinct, then

$$\mathbf{x}^{(r)\mathrm{T}}\mathbf{x}^{(s)} = \mathbf{0} \qquad (r \neq s) \tag{2.34}$$

By appropriate choice of the scalar multiplier, the eigenvector $\mathbf{x}^{(1)}$ can be *normalized* so that $\mathbf{x}^{(1)\mathrm{T}}\mathbf{x}^{(1)} = \mathbf{1}$. In general, we can normalize the eigenvectors so that

$$\mathbf{x}^{(r)\mathrm{T}}\mathbf{x}^{(s)} = \mathbf{1} \qquad (r = s) \tag{2.35}$$

Strictly speaking, the right-hand sides of (2.34) and (2.35) are 1×1 matrices, but for most purposes they may be treated as scalars. Now construct a 3×3 matrix $\mathbf{P}$ whose rows are the transposes of the normalized eigenvectors $\mathbf{x}^{(1)}$, $\mathbf{x}^{(2)}$, $\mathbf{x}^{(3)}$:

$$\mathbf{P}^{\mathrm{T}} = (\mathbf{x}^{(1)} \quad \mathbf{x}^{(2)} \quad \mathbf{x}^{(3)}) \tag{2.36}$$

Then it follows from (2.34) and (2.35) that $\mathbf{P}\mathbf{P}^{\mathrm{T}} = \mathbf{I}$, and so $\mathbf{P}$ is an orthogonal matrix. Also, using (2.28) and analogous relations for $\mathbf{x}^{(2)}$ and $\mathbf{x}^{(3)}$,

$$\mathbf{A}\mathbf{P}^{\mathrm{T}} = (\mathbf{A}\mathbf{x}^{(1)} \quad \mathbf{A}\mathbf{x}^{(2)} \quad \mathbf{A}\mathbf{x}^{(3)}) = (\lambda_1\mathbf{x}^{(1)} \quad \lambda_2\mathbf{x}^{(2)} \quad \lambda_3\mathbf{x}^{(3)}) \tag{2.37}$$

and hence, from (2.35), (2.36) and (2.37),

$$\mathbf{P}\mathbf{A}\mathbf{P}^{\mathrm{T}} = \begin{pmatrix} \lambda_1 & 0 & 0 \\ 0 & \lambda_2 & 0 \\ 0 & 0 & \lambda_3 \end{pmatrix} \tag{2.38}$$

Thus $\mathbf{P}\mathbf{A}\mathbf{P}^{\mathrm{T}}$ is a diagonal matrix with the eigenvalues of $\mathbf{A}$ as the elements on its leading diagonal.

It can be shown that if $\mathbf{A}$ is symmetric and $\lambda_1=\lambda_2\neq\lambda_3$, then the normalized eigenvector $\mathbf{x}^{(3)}$ is uniquely determined and $\mathbf{x}^{(1)}$ and $\mathbf{x}^{(2)}$ may be any two column matrices orthogonal to $\mathbf{x}^{(3)}$. If $\mathbf{x}^{(1)}$ and $\mathbf{x}^{(2)}$ are chosen to be mutually orthogonal then the results (2.33)–(2.38) remain valid. If $\lambda_1=\lambda_2=\lambda_3$, then $\mathbf{A}$ is diagonal. Any column matrix with at least one non-zero element is an eigenvector and the results remain true, though trivial, if $\mathbf{x}^{(1)}$, $\mathbf{x}^{(2)}$ and $\mathbf{x}^{(3)}$ are chosen as any three mutually orthogonal normalized column matrices.

From (2.23) it follows that

$$\mathbf{A}^2\mathbf{x}=\lambda\mathbf{A}\mathbf{x}=\lambda^2\mathbf{x} \tag{2.39}$$

Hence if λ is an eigenvalue of $\mathbf{A}$, and $\mathbf{x}$ is the corresponding eigenvector, then λ^2 is an eigenvalue of $\mathbf{A}^2$, and $\mathbf{x}$ is the corresponding eigenvector. More generally, λ^n is an eigenvalue of $\mathbf{A}^n$, and $\mathbf{x}$ is the corresponding eigenvector. If $\mathbf{A}$ is non-singular, this result holds for negative as well as for positive integers n.

2.4 The Cayley–Hamilton theorem

From (2.38), we see that

$$\operatorname{tr}\mathbf{PAP}^{\mathrm{T}}=\lambda_1+\lambda_2+\lambda_3, \qquad \operatorname{tr}(\mathbf{PAP}^{\mathrm{T}})^2=\lambda_1^2+\lambda_2^2+\lambda_3^2$$

Now since $\mathbf{P}$ is orthogonal, it follows from (2.15) that

$$\begin{aligned}\operatorname{tr}\mathbf{PAP}^{\mathrm{T}}&=P_{ij}A_{jk}P_{ik}=\delta_{jk}A_{jk}=A_{kk}=\operatorname{tr}\mathbf{A}\\ \operatorname{tr}(\mathbf{PAP}^{\mathrm{T}})^2&=\operatorname{tr}\mathbf{PAP}^{\mathrm{T}}\mathbf{PAP}^{\mathrm{T}}=\operatorname{tr}\mathbf{PAIAP}^{\mathrm{T}}\\ &=\operatorname{tr}\mathbf{PA}^2\mathbf{P}^{\mathrm{T}}=P_{ij}A_{jp}A_{pk}P_{ik}=\delta_{jk}A_{jp}A_{pk}=A_{kp}A_{pk}=\operatorname{tr}\mathbf{A}^2\end{aligned}$$

Hence

$$\lambda_1+\lambda_2+\lambda_3=\operatorname{tr}\mathbf{A}, \qquad \lambda_1^2+\lambda_2^2+\lambda_3^2=\operatorname{tr}\mathbf{A}^2 \tag{2.40}$$

From (2.25) and (2.26)

$$\lambda^3-(\lambda_1+\lambda_2+\lambda_3)\lambda^2+(\lambda_2\lambda_3+\lambda_3\lambda_1+\lambda_1\lambda_2)\lambda-\lambda_1\lambda_2\lambda_3=0$$

Hence, from (2.27) and (2.40), the characteristic equation can be expressed in the form

$$\lambda^3-\lambda^2\operatorname{tr}\mathbf{A}+\tfrac{1}{2}\lambda\{(\operatorname{tr}\mathbf{A})^2-\operatorname{tr}\mathbf{A}^2\}-\det\mathbf{A}=0 \tag{2.41}$$

The Cayley–Hamilton theorem states that a square matrix satisfies its own characteristic equation; thus for any 3×3 matrix $\mathbf{A}$,

$$\mathbf{A}^3-\mathbf{A}^2\,\mathrm{tr}\,\mathbf{A}+\tfrac{1}{2}\mathbf{A}\{(\mathrm{tr}\,\mathbf{A})^2-\mathrm{tr}\,\mathbf{A}^2\}-\mathbf{I}\det\mathbf{A}=\mathbf{0} \tag{2.42}$$

The theorem may be proved in several ways. Proofs will be found in standard algebra texts.

2.5 The polar decomposition theorem

A matrix $\mathbf{A}$ is *positive definite* if $\mathbf{x}^{\mathrm{T}}\mathbf{A}\mathbf{x}$ is positive for all non-zero values of the column matrix $\mathbf{x}$. A necessary and sufficient condition for $\mathbf{A}$ to be positive definite is that the eigenvalues of $\mathbf{A}$ are all positive.

The polar decomposition theorem states that a non-singular square matrix $\mathbf{F}$ can be decomposed, uniquely, into either of the products

$$\mathbf{F}=\mathbf{RU},\qquad \mathbf{F}=\mathbf{VR} \tag{2.43}$$

where $\mathbf{R}$ is an orthogonal matrix and $\mathbf{U}$ and $\mathbf{V}$ are positive definite symmetric matrices. We outline the proof for 3×3 matrices, which is the case we require. The generalization to $n\times n$ matrices is straightforward.

Let $\mathbf{C}=\mathbf{F}^{\mathrm{T}}\mathbf{F}$, and let $\bar{\mathbf{x}}=\mathbf{F}\mathbf{x}$. Then $\mathbf{C}$ is symmetric, and also

$$\mathbf{x}^{\mathrm{T}}\mathbf{C}\mathbf{x}=\mathbf{x}^{\mathrm{T}}\mathbf{F}^{\mathrm{T}}\mathbf{F}\mathbf{x}=\bar{\mathbf{x}}^{\mathrm{T}}\,\bar{\mathbf{x}}$$

But $\bar{\mathbf{x}}^{\mathrm{T}}\bar{\mathbf{x}}$ is a sum of squares, and so is positive for all non-zero column matrices $\bar{\mathbf{x}}$, and hence $\mathbf{x}^{\mathrm{T}}\mathbf{C}\mathbf{x}$ is positive for all non-zero $\mathbf{x}$. Thus $\mathbf{C}$ is positive definite, and has positive eigenvalues; we denote these by λ_1^2, λ_2^2, λ_3^2, where, without loss of generality, λ_1, λ_2 and λ_3 are positive. By the results of Section 2.3, if $\mathbf{P}^{\mathrm{T}}$ denotes the matrix whose columns are the normalized eigenvectors of $\mathbf{C}$, then $\mathbf{P}$ is orthogonal and

$$\mathbf{PCP}^{\mathrm{T}}=\begin{pmatrix}\lambda_1^2 & 0 & 0\\ 0 & \lambda_2^2 & 0\\ 0 & 0 & \lambda_3^2\end{pmatrix}$$

We define

$$\mathbf{U}=\mathbf{P}^{\mathrm{T}}\begin{pmatrix}\lambda_1 & 0 & 0\\ 0 & \lambda_2 & 0\\ 0 & 0 & \lambda_3\end{pmatrix}\mathbf{P} \tag{2.44}$$

Then $\mathbf{U}$ is symmetric and positive definite, and also, since $\mathbf{P}$ is orthogonal,

$$\mathbf{U}^2=\mathbf{P}^{\mathrm{T}}\begin{pmatrix}\lambda_1^2 & 0 & 0\\ 0 & \lambda_2^2 & 0\\ 0 & 0 & \lambda_3^2\end{pmatrix}\mathbf{P}=\mathbf{C} \tag{2.45}$$

We further define $\mathbf{R}=\mathbf{F}\mathbf{U}^{-1}$. Then in order to prove the existence of the first decomposition it is only necessary to show that $\mathbf{R}$ is orthogonal. Now from (2.43) and (2.45),

$$\mathbf{R}^{\mathrm{T}}\mathbf{R}=\mathbf{U}^{-1}\mathbf{F}^{\mathrm{T}}\mathbf{F}\mathbf{U}^{-1}=\mathbf{U}^{-1}\mathbf{C}\mathbf{U}^{-1}=\mathbf{U}^{-1}\mathbf{U}^2\mathbf{U}^{-1}=\mathbf{I}$$

and so $\mathbf{R}$ is indeed orthogonal. The matrix $\mathbf{V}$ is then defined by $\mathbf{V}=\mathbf{R}\mathbf{U}\mathbf{R}^{\mathrm{T}}$.

To prove uniqueness, suppose there exists another decomposition $\mathbf{F}=\mathbf{R}_1\mathbf{U}_1$, where $\mathbf{R}_1$ is orthogonal and $\mathbf{U}_1$ is positive definite. Then $\mathbf{U}_1^2=\mathbf{C}$, and

$$\mathbf{P}\mathbf{U}_1^2\mathbf{P}^{\mathrm{T}}=(\mathbf{P}\mathbf{U}_1\mathbf{P}^{\mathrm{T}})(\mathbf{P}\mathbf{U}_1\mathbf{P}^{\mathrm{T}})=\begin{pmatrix}\lambda_1^2 & 0 & 0\\ 0 & \lambda_2^2 & 0\\ 0 & 0 & \lambda_3^2\end{pmatrix}$$

Hence

$$\mathbf{P}\mathbf{U}_1\mathbf{P}^{\mathrm{T}}=\begin{pmatrix}\pm\lambda_1 & 0 & 0\\ 0 & \pm\lambda_2 & 0\\ 0 & 0 & \pm\lambda_3\end{pmatrix}\qquad \mathbf{U}_1=\mathbf{P}^{\mathrm{T}}\begin{pmatrix}\pm\lambda_1 & 0 & 0\\ 0 & \pm\lambda_2 & 0\\ 0 & 0 & \pm\lambda_3\end{pmatrix}\mathbf{P}$$

However, the only one of these matrices $\mathbf{U}_1$ which is positive definite is the one in which the positive signs are taken. Hence $\mathbf{U}_1=\mathbf{U}$. The uniqueness of $\mathbf{R}$ and $\mathbf{V}$ then follows from their definitions.

The above proof proceeds by constructing the matrices $\mathbf{U}$, $\mathbf{R}$ and $\mathbf{V}$ which correspond to a given matrix $\mathbf{F}$. Thus, in principle it gives a method of determining $\mathbf{U}$, $\mathbf{R}$ and $\mathbf{V}$. In practice, the calculations are cumbersome, even for a 3×3 matrix $\mathbf{F}$. Fortunately, for applications in continuum mechanics it is usually sufficient to know that the unique decompositions exist, and it is not often necessary to carry them out explicitly.

3

Vectors and cartesian tensors

3.1 Vectors

We assume familiarity with basic vector algebra and analysis. In the first part of this chapter we define the notation and summarize some of the more important results so that they are available for future reference.

We consider vectors in three-dimensional Euclidean space. Such vectors will (with a few exceptions which will be noted as they occur) be denoted by lower-case bold-face italic letters ($\boldsymbol{a}$, $\boldsymbol{b}$, $\boldsymbol{x}$, etc.). We make a distinction between column matrices, which are purely algebraic quantities introduced in Chapter 2, and vectors, which represent physical quantities such as displacement, velocity, acceleration, force, momentum, and so on. This distinction is reflected in our use of roman bold-face type for column matrices and italic bold-face type for vectors.

The characteristic properties of a vector are: (a) a vector requires a magnitude and a direction for its complete specification, and (b) two vectors are compounded in accordance with the parallelogram law. Thus two vectors $\boldsymbol{a}$ and $\boldsymbol{b}$ may be represented in magnitude and direction by two lines in space, and if these two lines are taken to be adjacent sides of a parallelogram, the vector sum $\boldsymbol{a}+\boldsymbol{b}$ is represented in magnitude and direction by the diagonal of the parallelogram which passes through the point of intersection of the two lines.

Suppose there is set up a system of rectangular right-handed cartesian coordinates with origin O. Let $\boldsymbol{e}_1$, $\boldsymbol{e}_2$, $\boldsymbol{e}_3$ denote vectors of unit magnitude in the directions of the three coordinate axes. Then $\boldsymbol{e}_1$, $\boldsymbol{e}_2$, $\boldsymbol{e}_3$ are called *base vectors* of the coordinate system. By virtue of the parallelogram addition law, a vector $\boldsymbol{a}$ can be expressed as a vector sum of three such unit vectors directed in the three coordinate directions. Thus

$$\boldsymbol{a}=a_1\boldsymbol{e}_1+a_2\boldsymbol{e}_2+a_3\boldsymbol{e}_3=a_i\boldsymbol{e}_i \qquad (3.1)$$

where in the last expression (and in future, whenever it is convenient) the summation convention is employed. The quantities a_i $(i=1, 2, 3)$ are the *components* of $\boldsymbol{a}$ in the specified coordinate system; they are related to the magnitude a of $\boldsymbol{a}$ by

$$a^2 = a_1^2 + a_2^2 + a_3^2 = a_i a_i \tag{3.2}$$

In particular, a vector may be the position vector $\boldsymbol{x}$ of a point P relative to O. Then the components x_1, x_2, x_3 of $\boldsymbol{x}$ are the *coordinates* of P in the given coordinate system, and the magnitude of $\boldsymbol{x}$ is the length OP.

The scalar product $\boldsymbol{a} \cdot \boldsymbol{b}$ of the two vectors $\boldsymbol{a}$, $\boldsymbol{b}$ with respective magnitudes a, b, whose directions are separated by an angle θ, is the scalar quantity

$$\boldsymbol{a} \cdot \boldsymbol{b} = ab \cos\theta = a_1 b_1 + a_2 b_2 + a_3 b_3 = a_i b_i \tag{3.3}$$

If $\boldsymbol{a}$ and $\boldsymbol{b}$ are parallel, then $\boldsymbol{a} \cdot \boldsymbol{b} = ab$, and if $\boldsymbol{a}$ and $\boldsymbol{b}$ are at right angles, $\boldsymbol{a} \cdot \boldsymbol{b} = 0$. In particular,

$$\boldsymbol{e}_i \cdot \boldsymbol{e}_j = \begin{cases} 0 & \text{if } i \neq j \\ 1 & \text{if } i = j \end{cases}$$

That is,

$$\boldsymbol{e}_i \cdot \boldsymbol{e}_j = \delta_{ij} \tag{3.4}$$

The vector product $\boldsymbol{a} \times \boldsymbol{b}$ of $\boldsymbol{a}$ and $\boldsymbol{b}$ is a vector whose direction is normal to the plane of $\boldsymbol{a}$ and $\boldsymbol{b}$, in the sense of a right-handed screw rotating from $\boldsymbol{a}$ to $\boldsymbol{b}$, and whose magnitude is $ab \sin\theta$. In terms of components, $\boldsymbol{a} \times \boldsymbol{b}$ can conveniently be written as

$$\boldsymbol{a} \times \boldsymbol{b} = \begin{vmatrix} \boldsymbol{e}_1 & \boldsymbol{e}_2 & \boldsymbol{e}_3 \\ a_1 & a_2 & a_3 \\ b_1 & b_2 & b_3 \end{vmatrix} \tag{3.5}$$

where it is understood that the determinant expansion is to be by the first row. By using the alternating symbol e_{ijk}, (3.5) can be written as

$$\boldsymbol{a} \times \boldsymbol{b} = e_{ijk} \boldsymbol{e}_i a_j b_k \tag{3.6}$$

The triple scalar product $(\boldsymbol{a} \times \boldsymbol{b}) \cdot \boldsymbol{c}$ is given in components as

$$(\boldsymbol{a} \times \boldsymbol{b}) \cdot \boldsymbol{c} = \begin{vmatrix} a_1 & a_2 & a_3 \\ b_1 & b_2 & b_3 \\ c_1 & c_2 & c_3 \end{vmatrix} = e_{ijk} a_i b_j c_k \tag{3.7}$$

3.2 Coordinate transformation

A vector is a quantity which is independent of any coordinate system. If a coordinate system is introduced the vector may be represented by its components in that system, but the same vector will have different components in different coordinate systems. Sometimes the components of a vector in a given coordinate system may conveniently be written as a column matrix, but this matrix only specifies the vector if the coordinate system is also specified.

Suppose the coordinate system is translated but not rotated, so that the new origin is O', where O' has position vector $\boldsymbol{x}_0$ relative to O. Then the position vector $\boldsymbol{x}'$ of P relative to O' is

$$\boldsymbol{x}' = \boldsymbol{x} - \boldsymbol{x}_0.$$

In a translation without rotation, the base vectors $\boldsymbol{e}_1$, $\boldsymbol{e}_2$, $\boldsymbol{e}_3$ are unchanged, and so the components a_i of a vector $\boldsymbol{a}$ are the same in the system with origin O' as they were in the system with origin O.

Now introduce a new rectangular right-handed cartesian coordinate system with the same origin O as the original system and base vectors $\bar{\boldsymbol{e}}_1$, $\bar{\boldsymbol{e}}_2$, $\bar{\boldsymbol{e}}_3$. The new system may be regarded as having been derived from the old by a rigid rotation of the triad of coordinate axes about O. Let a vector $\boldsymbol{a}$ have components a_i in the original coordinate system and components $\bar{a}_i$ in the new system. Thus

$$\boldsymbol{a} = a_i \boldsymbol{e}_i = \bar{a}_i \bar{\boldsymbol{e}}_i \tag{3.8}$$

Now denote by M_{ij} the cosine of the angle between $\bar{\boldsymbol{e}}_i$ and $\boldsymbol{e}_j$, so that

$$M_{ij} = \bar{\boldsymbol{e}}_i \cdot \boldsymbol{e}_j \tag{3.9}$$

Then M_{ij} $(i, j = 1, 2, 3)$ are the direction cosines of $\bar{\boldsymbol{e}}_i$ relative to the first coordinate system, or, equivalently, M_{ij} are the components of $\bar{\boldsymbol{e}}_i$ in the first system. Thus

$$\bar{\boldsymbol{e}}_i = M_{ij} \boldsymbol{e}_j \tag{3.10}$$

It is geometrically evident that the nine quantities M_{ij} are not independent. In fact, since $\bar{\boldsymbol{e}}_i$ are mutually orthogonal unit vectors, we have, as in (3.4) $\bar{\boldsymbol{e}}_i \cdot \bar{\boldsymbol{e}}_j = \delta_{ij}$. However, from (3.4) and (3.10),

$$\bar{\boldsymbol{e}}_i \cdot \bar{\boldsymbol{e}}_j = M_{ir} \boldsymbol{e}_r \cdot M_{js} \boldsymbol{e}_s = M_{ir} M_{js} \boldsymbol{e}_r \cdot \boldsymbol{e}_s = M_{ir} M_{js} \delta_{rs} = M_{ir} M_{jr}$$

Hence

$$M_{ir}M_{jr} = \delta_{ij} \tag{3.11}$$

Since $\delta_{ij} = \delta_{ji}$, (3.11) represents a set of six relations between the nine quantities M_{ij}. Now regard M_{ij} as the elements of a square matrix $\mathbf{M}$. Then (3.11) is equivalent to the statement

$$\mathbf{M}\mathbf{M}^{\mathrm{T}} = \mathbf{I} \tag{3.12}$$

Thus $\mathbf{M} = (M_{ij})$ is an orthogonal matrix; that is, the matrix which determines the new base vectors in terms of the old base vectors is an orthogonal matrix. For a transformation from one right-handed system to another right-handed system, $\mathbf{M}$ is a proper orthogonal matrix. The rows of $\mathbf{M}$ are the direction cosines of $\bar{\boldsymbol{e}}_i$ in the first coordinate system.

Since $\mathbf{M}$ is orthogonal, the reciprocal relation to (3.10) is

$$\boldsymbol{e}_j = M_{ij}\bar{\boldsymbol{e}}_i, \tag{3.13}$$

and so the columns of $\mathbf{M}$ are the direction cosines of the $\boldsymbol{e}_j$ in the coordinate system with base vectors $\bar{\boldsymbol{e}}_i$.

Now from (3.8) and (3.13),

$$\bar{a}_i\bar{\boldsymbol{e}}_i = a_j\boldsymbol{e}_j = a_jM_{ij}\bar{\boldsymbol{e}}_i$$

Thus

$$\bar{a}_i = M_{ij}a_j \tag{3.14}$$

This gives the new components $\bar{a}_i$ of $\boldsymbol{a}$ in terms of its old components a_j, and the elements of the orthogonal matrix $\mathbf{M}$ which determines the new base vectors in terms of the old. Similarly, from (3.8) and (3.10)

$$a_i = M_{ji}\bar{a}_j \tag{3.15}$$

In particular, if $\boldsymbol{a}$ is the position vector $\boldsymbol{x}$ of the point P relative to the origin O, then

$$\bar{x}_i = M_{ij}x_j, \qquad x_i = M_{ji}\bar{x}_j \tag{3.16}$$

where x_i and $\bar{x}_i$ are the coordinates of the point P in the first and second coordinate systems respectively.

The transformation law, (3.14) and (3.15), is a consequence of the parallelogram law of addition of vectors, and can be shown to be equivalent to this law. Thus a vector can be defined to be a quantity with magnitude and direction which: (a) compounds according to the parallelogram law, or equivalently, (b) can be represented by a set of components which transform as (3.14) under a rotation of the coordinate system.

In the foregoing discussion we have admitted only rotations of the coordinate system, so that $\mathbf{M}$ is a proper orthogonal matrix ($\det \mathbf{M} = 1$). If we also consider transformations from a right-handed to a left-handed coordinate system, for which $\mathbf{M}$ is an improper orthogonal matrix ($\det \mathbf{M} = -1$), then it becomes necessary to distinguish between vectors, whose components transform according to (3.14), and *pseudo-vectors*, whose components transform according to the rule

$$\bar{a}_i = (\det \mathbf{M}) M_{ij} a_j \tag{3.17}$$

Examples of pseudo-vectors are the vector product $\boldsymbol{a} \times \boldsymbol{b}$ of two vectors $\boldsymbol{a}$ and $\boldsymbol{b}$, the angular velocity vector, the infinitesimal rotation vector (Section 6.7) and the vorticity vector (Section 6.9). The distinction between vectors and pseudo-vectors only arises if left-handed coordinate systems are introduced, and it will not be of importance in this book.

It is evident from the definition of the scalar product $\boldsymbol{a} \cdot \boldsymbol{b}$ that its value must be independent of the choice of the coordinate system. To confirm this we observe from (3.14) that

$$\boldsymbol{a} \cdot \boldsymbol{b} = \bar{a}_i \bar{b}_i = M_{ij} a_j M_{ik} b_k = \delta_{jk} a_j b_k = a_j b_j = a_i b_i \tag{3.18}$$

A quantity such as $a_i b_i$, whose value is independent of the coordinate system to which the components are referred, is an *invariant* of the vectors $\boldsymbol{a}$ and $\boldsymbol{b}$.

As the vector product is also defined geometrically, it must have a similar invariance property. In fact, from (2.22), (3.10) and (3.14) we have

$$\begin{aligned} e_{ijk} \bar{\boldsymbol{e}}_i \bar{a}_j \bar{b}_k &= e_{ijk} M_{ip} M_{jq} M_{kr} \boldsymbol{e}_p a_q b_r \\ &= e_{pqr} (\det \mathbf{M}) \boldsymbol{e}_p a_q b_r \\ &= e_{pqr} \boldsymbol{e}_p a_q b_r = e_{ijk} \boldsymbol{e}_i a_j b_k \end{aligned} \tag{3.19}$$

provided that $\det \mathbf{M} = +1$.

The reader will observe the advantages of using the summation convention in equations such as (3.18) and (3.19). Not only does this notation allow lengthy sums to be expressed concisely (for example, the third expression in (3.18) represents a sum of 27 terms) but it also reveals the structure of these complicated expressions and suggests the ways in which they may be simplified.

3.3 The dyadic product

There are some physical quantities, apart from quantities which can be expressed as scalar or vector products, which require the specification of two vectors for their description. For example, to describe the force acting on a surface it is necessary to know the magnitude and direction of the force and the orientation of the surface. Some quantities of this kind can be described by a *dyadic product*.

The dyadic product of two vectors $\boldsymbol{a}$ and $\boldsymbol{b}$ is written $\boldsymbol{a} \otimes \boldsymbol{b}$. It has the properties

$$\begin{gathered}(\alpha \boldsymbol{a}) \otimes \boldsymbol{b} = \boldsymbol{a} \otimes (\alpha \boldsymbol{b}) = \alpha(\boldsymbol{a} \otimes \boldsymbol{b}) \\ \boldsymbol{a} \otimes (\boldsymbol{b}+\boldsymbol{c}) = \boldsymbol{a} \otimes \boldsymbol{b} + \boldsymbol{a} \otimes \boldsymbol{c}, \qquad (\boldsymbol{b}+\boldsymbol{c}) \otimes \boldsymbol{a} = \boldsymbol{b} \otimes \boldsymbol{a} + \boldsymbol{c} \otimes \boldsymbol{a}\end{gathered} \tag{3.20}$$

where α is a scalar. It follows that in terms of the components of $\boldsymbol{a}$ and $\boldsymbol{b}$, $\boldsymbol{a} \otimes \boldsymbol{b}$ may be written

$$\boldsymbol{a} \otimes \boldsymbol{b} = (a_i \boldsymbol{e}_i) \otimes (b_j \boldsymbol{e}_j) = a_i b_j \boldsymbol{e}_i \otimes \boldsymbol{e}_j \tag{3.21}$$

We note that, in general, $\boldsymbol{a} \otimes \boldsymbol{b} \neq \boldsymbol{b} \otimes \boldsymbol{a}$. The form of (3.21) is independent of the choice of coordinate system, for

$$\begin{aligned}\bar{a}_i \bar{b}_j \bar{\boldsymbol{e}}_i \otimes \bar{\boldsymbol{e}}_j &= M_{ip} a_p M_{jq} b_q (M_{ir} \boldsymbol{e}_r) \otimes (M_{js} \boldsymbol{e}_s) \\ &= M_{ip} M_{ir} M_{jq} M_{js} a_p b_q \boldsymbol{e}_r \otimes \boldsymbol{e}_s \\ &= \delta_{pr} \delta_{qs} a_p b_q \boldsymbol{e}_r \otimes \boldsymbol{e}_s \\ &= a_r b_s \boldsymbol{e}_r \otimes \boldsymbol{e}_s \\ &= a_i b_j \boldsymbol{e}_i \otimes \boldsymbol{e}_j \end{aligned} \tag{3.22}$$

The dyadic products $\boldsymbol{e}_i \otimes \boldsymbol{e}_j$ of the base vectors $\boldsymbol{e}_i$ are called *unit dyads*.

In addition to (3.20), the essential property of a dyadic product is that it forms an *inner product* with a vector, as follows

$$(\boldsymbol{a} \otimes \boldsymbol{b}) \cdot \boldsymbol{c} = \boldsymbol{a}(\boldsymbol{b} \cdot \boldsymbol{c}), \qquad \boldsymbol{a} \cdot (\boldsymbol{b} \otimes \boldsymbol{c}) = (\boldsymbol{a} \cdot \boldsymbol{b})\boldsymbol{c} \tag{3.23}$$

Since there is no possibility of ambiguity, the brackets on the left-hand sides of (3.23) may be omitted and we can write

$$\boldsymbol{a} \otimes \boldsymbol{b} \cdot \boldsymbol{c} = \boldsymbol{a}(\boldsymbol{b} \cdot \boldsymbol{c}), \qquad \boldsymbol{a} \cdot \boldsymbol{b} \otimes \boldsymbol{c} = (\boldsymbol{a} \cdot \boldsymbol{b})\boldsymbol{c} \tag{3.24}$$

Hence (3.24) can be written in terms of components as

$$\boldsymbol{a} \otimes \boldsymbol{b} \cdot \boldsymbol{c} = b_j c_j a_k \boldsymbol{e}_k, \qquad \boldsymbol{a} \cdot \boldsymbol{b} \otimes \boldsymbol{c} = a_i b_i c_k \boldsymbol{e}_k \tag{3.25}$$

Formally, $\boldsymbol{a} \cdot \boldsymbol{b}$ may be interpreted as the scalar product even when $\boldsymbol{a}$ or $\boldsymbol{b}$ form part of a dyadic product.

The concept of a dyadic product can be extended to products of three or more vectors. For example, a *triadic product* of the vectors $\boldsymbol{a}$, $\boldsymbol{b}$ and $\boldsymbol{c}$ is written $\boldsymbol{a} \otimes \boldsymbol{b} \otimes \boldsymbol{c}$ and can be expressed in component form as $a_i b_j c_k \boldsymbol{e}_i \otimes \boldsymbol{e}_j \otimes \boldsymbol{e}_k$.

3.4 Cartesian tensors

We define a *second-order cartesian tensor* to be a linear combination of dyadic products. As a dyadic product is, by (3.21), itself a linear combination of unit dyads, a second-order cartesian tensor $\boldsymbol{A}$ can be expressed as a linear combination of unit dyads, so that it takes the form

$$\boldsymbol{A} = A_{ij} \boldsymbol{e}_i \otimes \boldsymbol{e}_j \tag{3.26}$$

As a rule, we shall use bold-face italic capitals to denote cartesian tensors of second (and higher) order. As the only tensors which will be considered in this book until Chapter 11 will be cartesian tensors, we shall omit the adjective 'cartesian'. In Chapters 3–10, the term 'tensor' means 'cartesian tensor'.

The coefficients A_{ij} are called the *components* of $\boldsymbol{A}$. (Wherever possible, tensor components will be denoted by the same letter, in italic capitals, as is used to denote the tensor itself.) By the manner of its definition, a tensor exists independently of any coordinate system. However, its components can only be specified after a coordinate system has been introduced, and the values of the components depend on the choice of the coordinate system. Suppose that in a new coordinate system, with base vectors $\bar{\boldsymbol{e}}_i$, $\boldsymbol{A}$ has components $\bar{A}_{ij}$. Then

$$\boldsymbol{A} = A_{ij} \boldsymbol{e}_i \otimes \boldsymbol{e}_j = \bar{A}_{ij} \bar{\boldsymbol{e}}_i \otimes \bar{\boldsymbol{e}}_j \tag{3.27}$$

However, from (3.13),

$$A_{ij} \boldsymbol{e}_i \otimes \boldsymbol{e}_j = A_{ij} M_{pi} M_{qj} \bar{\boldsymbol{e}}_p \otimes \bar{\boldsymbol{e}}_q = \bar{A}_{pq} \bar{\boldsymbol{e}}_p \otimes \bar{\boldsymbol{e}}_q$$

Hence

$$\bar{A}_{pq} = M_{pi} M_{qj} A_{ij} \tag{3.28}$$

This is the *transformation law* for components of second-order tensors. It depends on the composition rule (3.20) and can be shown to be equivalent to this rule. Thus (3.28) may be used to formulate an alternative definition of a second-order tensor. In order to identify a second-order tensor as such, it is sufficient to show that in any transformation from one rectangular cartesian coordinate system to another, the components transform according to (3.28). In continuum mechanics, tensors are usually recognized by the property that their components transform in this manner.

More generally, a cartesian tensor of order n can be expressed in components as

$$\mathbf{A} = \underbrace{A_{ij\ldots m}}_{n \text{ indices}} \underbrace{\mathbf{e}_i \otimes \mathbf{e}_j \otimes \ldots \otimes \mathbf{e}_m}_{n \text{ factors}} \tag{3.29}$$

and its components transform according to the rule

$$\bar{A}_{pq\ldots t} = M_{pi}M_{qj} \ldots M_{tm}A_{ij\ldots m} \tag{3.30}$$

Thus a vector can be interpreted as a tensor of order one. A scalar, which has a single component which is unchanged in a coordinate transformation, can be regarded as a tensor of order zero. Nearly all of the tensors we encounter in this book will be of order zero (scalars), one (vectors), or two.

The inverse relation to (3.28) is

$$A_{ij} = M_{pi}M_{qj}\bar{A}_{pq}, \tag{3.31}$$

and the inverse of (3.30) is

$$A_{ij\ldots m} = M_{pi}M_{qj} \ldots M_{tm}\bar{A}_{pq\ldots t} \tag{3.32}$$

Suppose that $\mathbf{A} = A_{ij}\mathbf{e}_i \otimes \mathbf{e}_j = \bar{A}_{pq}\bar{\mathbf{e}}_p \otimes \bar{\mathbf{e}}_q$ is a second-order tensor, and that $A_{ij} = A_{ji}$. Then from (3.28),

$$\bar{A}_{qp} = M_{qi}M_{pj}A_{ij} = M_{pj}M_{qi}A_{ji} = \bar{A}_{pq} \tag{3.33}$$

Thus the property of symmetry with respect to interchange of tensor component indices is preserved under coordinate transformations, and so is a property of the tensor $\mathbf{A}$. A tensor $\mathbf{A}$ whose components have the property $A_{ij} = A_{ji}$ (in any coordinate system) is a *symmetric second-order tensor.* Many of the second-order tensors which occur in continuum mechanics are symmetric.

Similarly, if $A_{ij} = -A_{ji}$, then $\bar{A}_{ij} = -\bar{A}_{ji}$, and $\mathbf{A}$ is an *anti-symmetric second-order tensor.*

Let us denote $A_{ij}^{\mathrm{T}} = A_{ji}$ and $\bar{A}_{pq}^{\mathrm{T}} = \bar{A}_{qp}$. Then from (3.28),

$$\bar{A}_{pq}^{\mathrm{T}} = \bar{A}_{qp} = M_{qj}M_{pi}A_{ji} = M_{pi}M_{qj}A_{ij}^{\mathrm{T}} \tag{3.34}$$

Hence the set of components A_{ji} also transform as the components of a second-order tensor. Thus from the tensor $\mathbf{A} = A_{ij}\boldsymbol{e}_i \otimes \boldsymbol{e}_j$ we can form a new tensor $A_{ji}\boldsymbol{e}_i \otimes \boldsymbol{e}_j$ which we denote by $\boldsymbol{A}^{\mathrm{T}}$ and call the *transpose* of $\mathbf{A}$. The tensor $\mathbf{A} + \boldsymbol{A}^{\mathrm{T}}$ is symmetric and the tensor $\boldsymbol{A} - \boldsymbol{A}^{\mathrm{T}}$ is anti-symmetric. Since

$$\boldsymbol{A} = \tfrac{1}{2}(\boldsymbol{A} + \boldsymbol{A}^{\mathrm{T}}) + \tfrac{1}{2}(\boldsymbol{A} - \boldsymbol{A}^{\mathrm{T}}) \tag{3.35}$$

any second-order tensor can be decomposed into the sum of a symmetric and an anti-symmetric tensor, and this decomposition is unique.

3.5 Isotropic tensors

The tensor $\mathbf{I} = \delta_{ij}\boldsymbol{e}_i \otimes \boldsymbol{e}_j$ is called the *unit tensor*. In terms of another set of base vectors $\bar{\boldsymbol{e}}_i$, we have, from (3.13),

$$\begin{aligned} \boldsymbol{I} &= \delta_{ij}M_{ri}M_{sj}\bar{\boldsymbol{e}}_r \otimes \bar{\boldsymbol{e}}_s \\ &= M_{ri}M_{si}\bar{\boldsymbol{e}}_r \otimes \bar{\boldsymbol{e}}_s = \delta_{rs}\bar{\boldsymbol{e}}_r \otimes \bar{\boldsymbol{e}}_s = \delta_{ij}\bar{\boldsymbol{e}}_i \otimes \bar{\boldsymbol{e}}_j \end{aligned}$$

Thus the tensor $\boldsymbol{I}$ has the property that its components are δ_{ij} in any coordinate system. A tensor whose components are the same in any coordinate system is called an *isotropic tensor*. It can be shown that the only isotropic tensors of order two are of the form $p\boldsymbol{I}$, where p is a scalar. Such tensors are sometimes called *spherical tensors*.

Similarly, it can be verified that the *alternating tensor*

$$e_{ijk}\boldsymbol{e}_i \otimes \boldsymbol{e}_j \otimes \boldsymbol{e}_k \tag{3.36}$$

is an isotropic tensor of order three, provided that only coordinate transformations which correspond to proper orthogonal matrices (that is, rotations) are allowed. Any third-order isotropic tensor is a multiple of (3.36). There are three linearly independent fourth-order isotropic tensors, which may be taken to be

$$\delta_{ij}\delta_{kl}\boldsymbol{e}_i \otimes \boldsymbol{e}_j \otimes \boldsymbol{e}_k \otimes \boldsymbol{e}_l, \qquad \delta_{ik}\delta_{jl}\boldsymbol{e}_i \otimes \boldsymbol{e}_j \otimes \boldsymbol{e}_k \otimes \boldsymbol{e}_l,$$

$$\delta_{il}\delta_{jk}\boldsymbol{e}_i \otimes \boldsymbol{e}_j \otimes \boldsymbol{e}_k \otimes \boldsymbol{e}_l$$

and so the most general fourth-order isotropic tensor has the form

$$(\lambda\delta_{ij}\delta_{kl} + \mu\delta_{ik}\delta_{jl} + \nu\delta_{il}\delta_{jk})\boldsymbol{e}_i \otimes \boldsymbol{e}_j \otimes \boldsymbol{e}_k \otimes \boldsymbol{e}_l, \quad (3.37)$$

where λ, μ and ν are scalars.

3.6 Multiplication of tensors

Let $\boldsymbol{a} = a_i\boldsymbol{e}_i$ and $\boldsymbol{B} = B_{ij}\boldsymbol{e}_i \otimes \boldsymbol{e}_j$ be a vector and a second-order tensor respectively, with respective components a_i and B_{ij} in a coordinate system with base vectors $\boldsymbol{e}_i$. Suppose that in a new system with base vectors $\bar{\boldsymbol{e}}_i = M_{ij}\boldsymbol{e}_j$, $\boldsymbol{a}$ and $\boldsymbol{B}$ have components $\bar{a}_i$ and $\bar{B}_{ij}$ respectively, so that

$$\bar{a}_i = M_{ip}a_p, \qquad \bar{B}_{ij} = M_{ir}M_{js}B_{rs}$$

In addition, let $C_{ijk} = a_iB_{jk}$, and consider the tensor

$$\boldsymbol{C} = C_{ijk}\boldsymbol{e}_i \otimes \boldsymbol{e}_j \otimes \boldsymbol{e}_k$$

The components of $\boldsymbol{C}$ referred to base vectors $\bar{\boldsymbol{e}}_i$ are $\bar{C}_{ijk}$, where

$$\bar{C}_{ijk} = M_{ip}M_{jr}M_{ks}C_{prs} = M_{ip}M_{jr}M_{ks}a_pB_{rs} = \bar{a}_i\bar{B}_{jk} \quad (3.38)$$

The tensor $\boldsymbol{C}$ is called the *outer product* of the vector $\boldsymbol{a}$ and the tensor $\boldsymbol{B}$ (in that order), and is written $\boldsymbol{a} \otimes \boldsymbol{B}$. Equation (3.38) shows that the components of $\boldsymbol{C}$ are related to those of $\boldsymbol{a}$ and $\boldsymbol{B}$ in the same way in any coordinate system.

Similarly, if $\boldsymbol{A}$ and $\boldsymbol{B}$ are second-order tensors with respective components A_{ij} and B_{ij} in the system with base vectors $\boldsymbol{e}_i$, then the outer product $\boldsymbol{D} = \boldsymbol{A} \otimes \boldsymbol{B}$ is the fourth-order tensor with components $D_{ijkl} = A_{ij}B_{kl}$ in this system, and under a coordinate transformation the components of $\boldsymbol{D}$ transform to $\bar{D}_{ijkl} = \bar{A}_{ij}\bar{B}_{kl}$.

Outer products of three or more tensors or vectors are formed in a similar way, and the extension to tensors of higher order is direct. The outer product of a tensor of order m with a tensor of order n is a tensor of order $m+n$ (vectors are regarded as tensors of order one). The dyadic product of two vectors is the outer product of those vectors.

Contraction. Now consider a third-order tensor $C_{ijk}\boldsymbol{e}_i \otimes \boldsymbol{e}_j \otimes \boldsymbol{e}_k$. The components C_{ijk} transform according to the rule

$$\bar{C}_{ijk} = M_{ip}M_{jr}M_{ks}C_{prs}$$

We now sum on the last two indices of $\bar{C}_{ijk}$; that is, we form the three sums

$$\bar{C}_{i11}+\bar{C}_{i22}+\bar{C}_{i33}=\bar{C}_{ijj} \qquad (i=1,2,3).$$

Formally, this is accomplished by setting the second and third indices of C_{ijk} equal to each other. Then

$$\bar{C}_{ijj}=M_{ip}M_{jr}M_{js}C_{prs}=M_{ip}\delta_{rs}C_{prs}=M_{ip}C_{prr} \tag{3.39}$$

Thus the components C_{prr} transform as the components of a vector. More generally, if $D_{ij\ldots p\ldots q\ldots rs}$ are components of a tensor of order n, and we sum on any pair of its indices so as to form, for example, $D_{ij\ldots p\ldots p\ldots rs}$, the resulting quantities are the components of a tensor of order $n-2$. This operation of reducing the order of a tensor by two by summing on a pair of indices is called *contraction* of the tensor. In particular, if A_{ij} are components of a second-order tensor, then A_{ii} is a scalar.

A contraction may be performed on indices of two tensors which are factors in an outer product. Thus if a_i are components of a vector $\boldsymbol{a}$, and B_{ij} are components of a second-order tensor $\boldsymbol{B}$, then a_iB_{ij} are components of a vector, and so are $B_{ij}a_j$. We call these vectors *inner products* of $\boldsymbol{a}$ and $\boldsymbol{B}$, and write

$$a_iB_{ij}\boldsymbol{e}_j=\boldsymbol{a}\cdot\boldsymbol{B}, \qquad B_{ij}a_j\boldsymbol{e}_i=\boldsymbol{B}\cdot\boldsymbol{a} \tag{3.40}$$

Note that $\boldsymbol{a}\cdot\boldsymbol{B}=\boldsymbol{B}\cdot\boldsymbol{a}$ only if $\boldsymbol{B}$ is a symmetric tensor.

Inner products of second- and higher-order tensors are formed in a similar way. Let $\boldsymbol{A}$ and $\boldsymbol{B}$ be second-order tensors with components A_{ij} and B_{ij} respectively. From them we can form various inner products, which are second-order tensors; for example,

$$\boldsymbol{A}\cdot\boldsymbol{B}=A_{ij}B_{jk}\boldsymbol{e}_i\otimes\boldsymbol{e}_k, \qquad \boldsymbol{A}^{\mathrm{T}}\cdot\boldsymbol{B}=A_{ji}B_{jk}\boldsymbol{e}_i\otimes\boldsymbol{e}_k \tag{3.41}$$

We note, for example, that

$$(\boldsymbol{A}\cdot\boldsymbol{B})^{\mathrm{T}}=\boldsymbol{B}^{\mathrm{T}}\cdot\boldsymbol{A}^{\mathrm{T}}$$

As a special case, the tensors $\boldsymbol{A}$ and $\boldsymbol{B}$ may be the same tensor. The tensor $\boldsymbol{A}\cdot\boldsymbol{A}$ is denoted by $\boldsymbol{A}^2$.

If there exists a tensor $\boldsymbol{A}^{-1}$ such that

$$\boldsymbol{A}\cdot\boldsymbol{A}^{-1}=\boldsymbol{I}, \qquad \boldsymbol{A}^{-1}\cdot\boldsymbol{A}=\boldsymbol{I} \tag{3.42}$$

then $\boldsymbol{A}^{-1}$ is called the *inverse tensor* to $\boldsymbol{A}$.

If the tensors $\boldsymbol{A}^{\mathrm{T}}$ and $\boldsymbol{A}^{-1}$ are equal, so that

$$\boldsymbol{A}^{\mathrm{T}} = \boldsymbol{A}^{-1} \tag{3.43}$$

then $\boldsymbol{A}$ is said to be an *orthogonal tensor.*

By using the polar decomposition theorem (Section 2.5), the components F_{ij} of a second-order tensor $\boldsymbol{F}$ can (provided that $\det(F_{ij}) \neq 0$) be decomposed uniquely in the forms

$$F_{ij} = R_{ik}U_{kj}, \qquad F_{ij} = V_{ik}R_{kj}$$

where R_{ik} are elements of an orthogonal matrix, and U_{ij} and V_{ij} are elements of positive definite symmetric matrices. We define the second-order tensors $\boldsymbol{R}$, $\boldsymbol{U}$ and $\boldsymbol{V}$ to be

$$\boldsymbol{R} = R_{ij}\boldsymbol{e}_i \otimes \boldsymbol{e}_j, \qquad \boldsymbol{U} = U_{ij}\boldsymbol{e}_i \otimes \boldsymbol{e}_j, \qquad \boldsymbol{V} = V_{ij}\boldsymbol{e}_i \otimes \boldsymbol{e}_j$$

Then $\boldsymbol{R}$ is an orthogonal tensor and $\boldsymbol{U}$ and $\boldsymbol{V}$ are symmetric tensors, and

$$\boldsymbol{R} \cdot \boldsymbol{U} = R_{ik}U_{kj}\boldsymbol{e}_i \otimes \boldsymbol{e}_j = F_{ij}\boldsymbol{e}_i \otimes \boldsymbol{e}_j = \boldsymbol{F}$$

and

$$\boldsymbol{V} \cdot \boldsymbol{R} = V_{ik}R_{kj}\boldsymbol{e}_i \otimes \boldsymbol{e}_j = F_{ij}\boldsymbol{e}_i \otimes \boldsymbol{e}_j = \boldsymbol{F}$$

Thus the tensor $\boldsymbol{F}$ can be decomposed into either of the inner products

$$\boldsymbol{F} = \boldsymbol{R} \cdot \boldsymbol{U}, \qquad \boldsymbol{F} = \boldsymbol{V} \cdot \boldsymbol{R} \tag{3.44}$$

3.7 Tensor and matrix notation

Relations between tensor quantities may be expressed either in *direct form*, as relations between scalars $\alpha, \beta, \ldots$, vectors $\boldsymbol{a}, \boldsymbol{b}, \ldots$, and tensors $\boldsymbol{A}, \boldsymbol{B}, \ldots$; or in *component form*, as relations between scalars $\alpha, \beta, \ldots$, vector components $a_i, b_i, \ldots$, and tensor components $A_{ij}, B_{ij}, \ldots$. The direct notation has the advantage that it emphasizes that physical statements are independent of the choice of the coordinate system. However, this advantage is not entirely lost when the component notation is used, because relations in component notation must be written in such a way that they preserve their form under coordinate transformations. The component form, used in conjunction with the summation convention, is often convenient for carrying out algebraic manipulations, and in considering specific problems it is always necessary at some stage to introduce a coordinate system and components. Some examples of the interchange between the different forms are given in Table 3.1. In this book we employ both notations, as convenient.

When it is necessary to transform components from one coordinate system to another, it is often convenient to introduce matrix notation. Suppose that $\boldsymbol{a}$ is a vector and $\boldsymbol{A}$ is a second-order tensor. Let $\boldsymbol{a}$ and $\boldsymbol{A}$ have components a_i and A_{ij} respectively in a coordinate system with base vectors $\boldsymbol{e}_i$, and components $\bar{a}_i$ and $\bar{A}_{ij}$ respectively in a coordinate system with base vectors $\bar{\boldsymbol{e}}_i$, where, as in (3.10), $\bar{\boldsymbol{e}}_i = M_{ij}\boldsymbol{e}_j$ and M_{ij} are elements of an orthogonal matrix $\mathbf{M}$. Then the transformation rules (3.14) and (3.28) for the components of $\boldsymbol{a}$ and $\boldsymbol{A}$ are

$$\bar{a}_i = M_{ij}a_j, \qquad \bar{A}_{ij} = M_{ip}M_{jq}A_{pq} \tag{3.45}$$

The components a_i and $\bar{a}_i$ may be arranged as the elements of two 3×1 column matrices $\mathbf{a}$ and $\bar{\mathbf{a}}$, thus

$$\mathbf{a} = (a_1 \quad a_2 \quad a_3)^{\mathrm{T}}, \qquad \bar{\mathbf{a}} = (\bar{a}_1 \quad \bar{a}_2 \quad \bar{a}_3)^{\mathrm{T}} \tag{3.46}$$

and the components A_{ij} and $\bar{A}_{ij}$ may be arranged as elements of two 3×3 matrices $\mathbf{A}$ and $\bar{\mathbf{A}}$, thus

$$\mathbf{A} = (A_{ij}), \qquad \bar{\mathbf{A}} = (\bar{A}_{ij}) \tag{3.47}$$

Then the transformation rules (3.45) may be written in matrix notation as

$$\bar{\mathbf{a}} = \mathbf{Ma}, \qquad \bar{\mathbf{A}} = \mathbf{MAM}^{\mathrm{T}}. \tag{3.48}$$

Since $\mathbf{M}$ is orthogonal, we immediately obtain the reciprocal relations

$$\mathbf{a} = \mathbf{M}^{\mathrm{T}}\bar{\mathbf{a}}, \qquad \mathbf{A} = \mathbf{M}^{\mathrm{T}}\bar{\mathbf{A}}\mathbf{M} \tag{3.49}$$

Matrix notation is also useful in carrying out algebraic manipulations which involve components of vector and tensor products. In Table 3.1 we list a number of examples of vector and tensor equations expressed in direct notation, component notation and matrix notation. In Table 3.1, α is a scalar, $\boldsymbol{a}$ and $\boldsymbol{b}$ are vectors with components a_i and b_i respectively, and $\boldsymbol{A}$, $\boldsymbol{B}$, $\boldsymbol{C}$, $\boldsymbol{D}$, are second-order tensors with components A_{ij}, B_{ij}, C_{ij}, D_{ij} respectively. Also $\mathbf{a}$ and $\mathbf{b}$ are 3×1 column matrices with elements a_i and b_i respectively, and $\mathbf{A}$, $\mathbf{B}$, $\mathbf{C}$, $\mathbf{D}$ are 3×3 matrices with elements A_{ij}, B_{ij}, C_{ij}, D_{ij} respectively.

Table 3.1 Examples of tensor and matrix notation

Direct tensor notation	Tensor component notation	Matrix notation
$\alpha = \boldsymbol{a} \cdot \boldsymbol{b}$	$\alpha = a_i b_i$	$(\alpha) = \mathbf{a}^{\mathrm{T}}\mathbf{b}$
$\boldsymbol{A} = \boldsymbol{a} \otimes \mathbf{b}$	$A_{ij} = a_i b_j$	$\mathbf{A} = \mathbf{a}\mathbf{b}^{\mathrm{T}}$
$\boldsymbol{b} = \boldsymbol{A} \cdot \boldsymbol{a}$	$b_i = A_{ij} a_j$	$\mathbf{b} = \mathbf{A}\mathbf{a}$
$\boldsymbol{b} = \boldsymbol{a} \cdot \boldsymbol{A}$	$b_j = a_i A_{ij}$	$\mathbf{b}^{\mathrm{T}} = \mathbf{a}^{\mathrm{T}}\mathbf{A}$
$\alpha = \boldsymbol{a} \cdot \boldsymbol{A} \cdot \boldsymbol{b}$	$\alpha = a_i A_{ij} b_j$	$(\alpha) = \mathbf{a}^{\mathrm{T}}\mathbf{A}\mathbf{b}$
$\boldsymbol{C} = \boldsymbol{A} \cdot \boldsymbol{B}$	$C_{ij} = A_{ik} B_{kj}$	$\mathbf{C} = \mathbf{A}\mathbf{B}$
$\boldsymbol{C} = \boldsymbol{A} \cdot \boldsymbol{B}^{\mathrm{T}}$	$C_{ij} = A_{ik} B_{jk}$	$\mathbf{C} = \mathbf{A}\mathbf{B}^{\mathrm{T}}$
$\boldsymbol{D} = \boldsymbol{A} \cdot \boldsymbol{B} \cdot \boldsymbol{C}$	$D_{ij} = A_{ik} B_{km} C_{mj}$	$\mathbf{D} = \mathbf{A}\mathbf{B}\mathbf{C}$

Since $\mathbf{A}\mathbf{A}^{-1} = \mathbf{A}^{-1}\mathbf{A} = \boldsymbol{I}$, it follows that if $\mathbf{A}$ is the matrix of components of $\boldsymbol{A}$, then $\mathbf{A}^{-1}$ is the matrix of components of $\boldsymbol{A}^{-1}$, in the same coordinate system. Hence the tensor $\boldsymbol{A}^{-1}$ exists only if $\det \mathbf{A} \neq 0$.

It is important not to confuse the vector $\boldsymbol{a}$ with the column matrix $\mathbf{a}$, nor the tensor $\boldsymbol{A}$ with the square matrix $\mathbf{A}$. In a given coordinate system, the matrix $\mathbf{a}$ serves to describe the vector $\boldsymbol{a}$. However, the vector is represented by different matrices in different coordinate systems, whereas the vector itself is independent of the coordinate system. Similarly, the matrix $\mathbf{A}$ describes the tensor $\boldsymbol{A}$ in a given coordinate system, but $\boldsymbol{A}$ has different matrix representations in different coordinates, although $\boldsymbol{A}$ itself is independent of the coordinate system.

3.8 Invariants of a second-order tensor

Let $\boldsymbol{A}$ be a second-order tensor with components A_{ij} in the coordinate system with base vectors $\boldsymbol{e}_i$ and components $\bar{A}_{ij}$ in the coordinate system with base vectors $\bar{\boldsymbol{e}}_i = M_{ij}\boldsymbol{e}_j$. Also, let $\mathbf{A} = (A_{ij})$, $\bar{\mathbf{A}} = (\bar{A}_{ij})$ and $\mathbf{M} = (M_{ij})$. Suppose that λ is an eigenvalue of $\bar{\mathbf{A}}$, so that

$$\det(\bar{\mathbf{A}} - \lambda\mathbf{I}) = 0$$

Then $\bar{\mathbf{A}} = \mathbf{M}\mathbf{A}\mathbf{M}^{\mathrm{T}}$ and $\mathbf{M}$ is an orthogonal matrix. Therefore,

$$\det\{\mathbf{M}(\mathbf{A} - \lambda\mathbf{I})\mathbf{M}^{\mathrm{T}}\} = 0$$

and hence

$$\det \mathbf{M} \det(\mathbf{A} - \lambda\mathbf{I}) \det \mathbf{M} = 0$$

However, since $\mathbf{M}$ is an orthogonal matrix, $(\det \mathbf{M})^2 = 1$, and so

$$\det(\mathbf{A} - \lambda\mathbf{I}) = 0$$

Hence λ is also an eigenvalue of $\boldsymbol{A}$. Thus the eigenvalues of the matrix of components of $\boldsymbol{A}$ are independent of the coordinate system to which these components are referred. The eigenvalues are intrinsic to the tensor $\boldsymbol{A}$; if $\boldsymbol{A}$ is symmetric they are real numbers (cf. Section 2.3) and they are then called the *principal components* or the *principal values* of $\boldsymbol{A}$. We denote the principal values of $\boldsymbol{A}$ by A_1, A_2 and A_3. If A_1, A_2 and A_3 are all positive, then $\boldsymbol{A}$ is a *positive definite* tensor.

Suppose that $\boldsymbol{A}$ is symmetric. If A_1, A_2 and A_3 are distinct, then the normalized eigenvectors $\mathbf{x}^{(1)}$, $\mathbf{x}^{(2)}$ and $\mathbf{x}^{(3)}$ of $\mathbf{A}$ are unique and mutually orthogonal, and

$$\mathbf{A}\mathbf{x}^{(i)} = A_i\mathbf{x}^{(i)} \qquad (i = 1, 2, 3;\ \text{no summation})$$

Also, since $\mathbf{M}$ is an orthogonal matrix, it follows that

$$\mathbf{M}\mathbf{A}\mathbf{M}^{\mathrm{T}}\mathbf{M}\mathbf{x}^{(i)} = \bar{\mathbf{A}}\mathbf{M}\mathbf{x}^{(i)} = A_i\mathbf{M}\mathbf{x}^{(i)} \qquad (\text{no summation})$$

Hence, if the vectors $\boldsymbol{x}_i$ are defined as

$$\boldsymbol{x}_i = x_j^{(i)}\boldsymbol{e}_j \tag{3.50}$$

then we have

$$\boldsymbol{A} \cdot \boldsymbol{x}_i = A_i\boldsymbol{x}_i \qquad (\text{no summation})$$

Let us refer $\boldsymbol{A}$ to a coordinate system in which $\boldsymbol{x}_i$ are the base vectors, so that we now identify $\bar{\boldsymbol{e}}_i$ with $\boldsymbol{x}_i$. Then, from (3.50), the matrix $\mathbf{P}$ of the transformation from coordinates with base vectors $\boldsymbol{e}_i$ to coordinates with base vectors $\boldsymbol{x}_i$ is (P_{ij}), where

$$P_{ij} = x_j^{(i)}, \qquad \mathbf{P}^{\mathrm{T}} = (\mathbf{x}^{(1)} \quad \mathbf{x}^{(2)} \quad \mathbf{x}^{(3)})$$

Therefore (cf. Section 2.3) from (2.38) and (3.48),

$$\bar{\mathbf{A}} = \mathbf{P}\mathbf{A}\mathbf{P}^{\mathrm{T}} = \begin{pmatrix} A_1 & 0 & 0 \\ 0 & A_2 & 0 \\ 0 & 0 & A_3 \end{pmatrix} \tag{3.51}$$

Thus there exists a coordinate system in which the matrix of components of a symmetric second-order tensor $\boldsymbol{A}$ is a diagonal matrix whose diagonal elements are the principal values of $\boldsymbol{A}$. This coordinate system has base vectors $\boldsymbol{x}_i$. Its axes are the *principal axes* of $\boldsymbol{A}$, and their directions are the *principal directions* of $\boldsymbol{A}$.

These results remain valid if A_1, A_2 and A_3 are not all distinct. If $A_1 = A_2 \neq A_3$, then the vector $\boldsymbol{x}_3$ is uniquely determined, and $\boldsymbol{x}_1$ and $\boldsymbol{x}_2$ may be taken to be any two unit vectors which are orthogonal to each other and to $\boldsymbol{x}_3$. If $A_1 = A_2 = A_3$, then the principal axes may be taken to be any three mutually orthogonal axes, and $\mathbf{A}$ is a spherical tensor.

If, for example, the principal axis determined by $\boldsymbol{x}_3$ coincides with the base vector $\boldsymbol{e}_3$, then $A_{13} = 0$, $A_{23} = 0$. Conversely, if $A_{13} = A_{23} = 0$, then the direction of $\boldsymbol{x}_3$ is a principal direction.

It follows from (2.39) that the principal values of $\mathbf{A}^2$ are A_1^2, A_2^2 and A_3^2. More generally, the principal values of $\mathbf{A}^n$ are A_1^n, A_2^n and A_3^n. This holds for negative as well as positive integers n provided that A_1, A_2 and A_3 are all non-zero. The principal axes of $\mathbf{A}^n$ coincide with those of $\mathbf{A}$.

It was emphasized above that the principal values of $\mathbf{A}$ are independent of the choice of the coordinate system; they are *invariants* of the tensor $\mathbf{A}$. Invariants play an important role in continuum mechanics. It can be shown that if $\mathbf{A}$ is symmetric then A_1, A_2 and A_3 are basic invariants in the sense that any invariant of $\mathbf{A}$ can be expressed in terms of them. In many applications it is more convenient to choose as the basic invariants three symmetric functions of A_1, A_2 and A_3 rather than the principal values themselves. Three such symmetric functions are

$$A_1 + A_2 + A_3, \qquad A_1^2 + A_2^2 + A_3^2, \qquad A_1^3 + A_2^3 + A_3^3 \tag{3.52}$$

These three quantities are clearly invariants and they are independent in the sense that no one of them can be expressed in terms of the other two.

The convenience of the set (3.52) results partly because they can be calculated from the tensor components in any coordinate system without going through the tedious calculation of A_1, A_2 and A_3. We see from (3.51) that

$$A_1 + A_2 + A_3 = \operatorname{tr} \bar{\mathbf{A}}.$$

However, since $\mathbf{P}$ is orthogonal,

$$\operatorname{tr} \bar{\mathbf{A}} = \bar{A}_{ii} = P_{ir} P_{is} A_{rs} = \delta_{rs} A_{rs} = A_{rr} = \operatorname{tr} \mathbf{A} \tag{3.53}$$

Thus the first of the invariants (3.52) is equal, in any coordinate system, to the trace of the matrix of components of $\mathbf{A}$. Similarly,

$$\begin{aligned} A_1^2 + A_2^2 + A_3^2 &= \operatorname{tr} \bar{\mathbf{A}}^2 = \bar{A}_{ik} \bar{A}_{ki} = P_{ip} P_{kq} A_{pq} P_{kr} P_{is} A_{rs} \\ &= \delta_{ps} \delta_{qr} A_{pq} A_{rs} = A_{pr} A_{rp} = \operatorname{tr} \mathbf{A}^2 \end{aligned} \tag{3.54}$$

and in a similar way it follows that

$$A_1^3+A_2^3+A_3^3=\operatorname{tr}\mathbf{A}^3$$

Since $\operatorname{tr}\mathbf{A}$ is independent of the choice of the coordinate system, we can without ambiguity define $\operatorname{tr}\mathbf{A}=\operatorname{tr}\boldsymbol{A}$. Similarly, we define $\operatorname{tr}\mathbf{A}^2=\operatorname{tr}\boldsymbol{A}^2$ and $\operatorname{tr}\mathbf{A}^3=\operatorname{tr}\boldsymbol{A}^3$, so that the set of invariants (3.52) may be expressed as

$$\{\operatorname{tr}\boldsymbol{A}, \operatorname{tr}\boldsymbol{A}^2, \operatorname{tr}\boldsymbol{A}^3\} \tag{3.55}$$

Only matrix multiplications are needed in order to calculate the set (3.55).

Another set of symmetric functions of A_1, A_2 and A_3 is $\{I_1, I_2, I_3\}$, where

$$I_1=A_1+A_2+A_3, \qquad I_2=A_2A_3+A_3A_1+A_1A_2, \qquad I_3=A_1A_2A_3 \tag{3.56}$$

These are clearly invariant quantities. I_2 can be expressed in terms of components of $\bar{\mathbf{A}}$ as follows:

$$\begin{aligned} I_2 &= \tfrac{1}{2}\{(A_1+A_2+A_3)^2-(A_1^2+A_2^2+A_3^2)\} \\ &= \tfrac{1}{2}\{(\operatorname{tr}\bar{\mathbf{A}})^2-\operatorname{tr}\bar{\mathbf{A}}^2\} \\ &= \tfrac{1}{2}\{(\operatorname{tr}\boldsymbol{A})^2-\operatorname{tr}\boldsymbol{A}^2\} \end{aligned}$$

For I_3 we have

$$\begin{aligned} I_3 &= \det\bar{\mathbf{A}}=\det(\mathbf{PAP}^{\mathrm{T}}) \\ &= \det\mathbf{P}\det\mathbf{A}\det\mathbf{P}^{\mathrm{T}} \\ &= \det\mathbf{A} \end{aligned}$$

Hence without ambiguity we may define $\det\mathbf{A}=\det\boldsymbol{A}=I_3$, and a set of three independent invariants of $\boldsymbol{A}$ (and the set usually used in practice) is $\{I_1, I_2, I_3\}$, where

$$I_1=\operatorname{tr}\boldsymbol{A}, \qquad I_2=\tfrac{1}{2}\{(\operatorname{tr}\boldsymbol{A})^2-\operatorname{tr}\boldsymbol{A}^2\}, \qquad I_3=\det\boldsymbol{A} \tag{3.57}$$

From (2.42) we see that the Cayley–Hamilton theorem for $\mathbf{A}$ can be expressed as

$$\mathbf{A}^3-I_1\mathbf{A}^2+I_2\mathbf{A}-I_3\mathbf{I}=0 \tag{3.58}$$

By taking the trace of (3.58), and remembering that $\operatorname{tr}I=3$, there follows an alternative expression for $I_3=\det\boldsymbol{A}$:

$$\begin{aligned} I_3 &= \tfrac{1}{3}\{\operatorname{tr}\boldsymbol{A}^3-I_1\operatorname{tr}\boldsymbol{A}^2+I_2\operatorname{tr}\boldsymbol{A}\} \\ &= \tfrac{1}{3}\{\operatorname{tr}\boldsymbol{A}^3-\tfrac{3}{2}\operatorname{tr}\boldsymbol{A}^2\operatorname{tr}\boldsymbol{A}+\tfrac{1}{2}(\operatorname{tr}\boldsymbol{A})^3\} \end{aligned} \tag{3.59}$$

3.9 Deviatoric tensors

The tensor

$$\boldsymbol{A}' = \boldsymbol{A} - \tfrac{1}{3}\boldsymbol{I} \operatorname{tr} \boldsymbol{A} \tag{3.60}$$

has the property that its first invariant, tr $\boldsymbol{A}'$, is zero. Thus, if $\boldsymbol{A}'$ is symmetric, it has only five independent components, and only two independent non-zero invariants. A tensor whose trace is zero is called a *deviatoric tensor* and $\boldsymbol{A}'$ is called the *deviator* of $\boldsymbol{A}$. It is sometimes useful in continuum mechanics to decompose a tensor into the sum of its deviator and a spherical tensor, as follows:

$$\boldsymbol{A} = \boldsymbol{A}' + \tfrac{1}{3}\boldsymbol{I} \operatorname{tr} \boldsymbol{A} \tag{3.61}$$

The two non-zero invariants of $\boldsymbol{A}'$ are

$$I_2' = -\tfrac{1}{2}\{(\operatorname{tr} \mathbf{A}')^2 - \operatorname{tr} \mathbf{A}'^2\}, \qquad I_3' = \det \mathbf{A}' = \tfrac{1}{3} \operatorname{tr} \mathbf{A}'^3 \tag{3.62}$$

After some manipulation it can be shown from (3.57) and (3.60) that

$$I_2' = -\tfrac{1}{3}I_1^2 + I_2, \qquad I_3' = I_3 - \tfrac{1}{3}I_1 I_2 + \tfrac{2}{27}I_1^3 \tag{3.63}$$

Thus I_2' and I_3' can be expressed in terms of I_1, I_2 and I_3. Alternatively, I_2 and I_3 can be expressed in terms of I_1. I_2' and I_3', and so $\{I_1, I_2, I_3'\}$ may be adopted as a set of basic invariants for $\boldsymbol{A}$ which is equivalent to the set $\{I_1, I_2, I_3\}$.

3.10 Vector and tensor calculus

We assume familiarity with elementary vector analysis, and give only a summary, without proof, of results which will be needed.

If $\phi(x_1, x_2, x_3)$ is a scalar function of the coordinates then

$$\operatorname{grad} \phi = \boldsymbol{\nabla}\phi = \boldsymbol{e}_1 \frac{\partial \phi}{\partial x_1} + \boldsymbol{e}_2 \frac{\partial \phi}{\partial x_2} + \boldsymbol{e}_3 \frac{\partial \phi}{\partial x_3} = \boldsymbol{e}_i \frac{\partial \phi}{\partial x_i} \tag{3.64}$$

is the gradient of ϕ and is a vector. grad ϕ is a vector whose direction is normal to a level surface $\phi(x_1, x_2, x_3) = \text{constant}$ and whose magnitude is the directional derivative of ϕ in the direction of this normal.

If $\boldsymbol{a}(x_1, x_2, x_3) = a_i(x_j)\boldsymbol{e}_i$ is a vector function of the coordinates then

$$\operatorname{div} \boldsymbol{a} = \boldsymbol{\nabla} \cdot \boldsymbol{a} = \frac{\partial a_1}{\partial x_1} + \frac{\partial a_2}{\partial x_2} + \frac{\partial a_3}{\partial x_3} = \frac{\partial a_i}{\partial x_i} \tag{3.65}$$

is the divergence of $\boldsymbol{a}$ and is a scalar. Also

$$\operatorname{curl} \boldsymbol{a} = \boldsymbol{\nabla} \times \boldsymbol{a} = \begin{vmatrix} \boldsymbol{e}_1 & \boldsymbol{e}_2 & \boldsymbol{e}_3 \\ \dfrac{\partial}{\partial x_1} & \dfrac{\partial}{\partial x_2} & \dfrac{\partial}{\partial x_3} \\ a_1 & a_2 & a_3 \end{vmatrix} = e_{ijk}\boldsymbol{e}_i \frac{\partial a_k}{\partial x_j} \tag{3.66}$$

is the curl of $\boldsymbol{a}$ and is a vector. In the symbolic determinant in (3.66) the expansion is to be carried out by the first row.

In continuum mechanics we make frequent use of the divergence theorem (or Gauss's theorem) which states that if the vector field $\boldsymbol{a}$ has continuous first-order partial derivatives at all points of a region $\mathcal{R}$ bounded by a surface $\mathcal{S}$, then

$$\iiint\limits_{\mathcal{R}} \operatorname{div} \boldsymbol{a} \, \mathrm{d}V = \iint\limits_{\mathcal{S}} \boldsymbol{a} \cdot \boldsymbol{n} \, \mathrm{d}S \tag{3.67}$$

where $\mathrm{d}V$ and $\mathrm{d}S$ are elements of volume and of surface area respectively, and $\boldsymbol{n}$ is the outward normal to $\mathcal{S}$. In terms of components, (3.67) takes the form

$$\iiint\limits_{\mathcal{R}} \frac{\partial a_i}{\partial x_i} \, \mathrm{d}V = \iint\limits_{\mathcal{S}} a_i n_i \, \mathrm{d}S \tag{3.68}$$

The divergence theorem can also be applied to tensors. For example, if $\boldsymbol{A}$ is a second-order tensor with components A_{ij} then

$$\iiint\limits_{\mathcal{R}} \frac{\partial A_{ij}}{\partial x_i} \, \mathrm{d}V = \iint\limits_{\mathcal{S}} A_{ij} n_i \, \mathrm{d}S \tag{3.69}$$

and analogous results hold for tensors of higher order.

4 Particle kinematics

4.1 Bodies and their configurations

Kinematics is the study of motion, without regard to the forces which produce it. In this chapter we discuss the motion of individual particles (although these particles may form part of a continuous body) without reference to the motion of neighbouring particles. The deformation, or change of shape, of a body depends on the motion of each particle relative to its neighbours, and will be analysed in Chapters 6 and 9.

We introduce a fixed rectangular cartesian coordinate system with origin O and base vectors $\boldsymbol{e}_i$. Throughout Chapters 4 to 10 all motion will be motion relative to this fixed frame of reference and, unless otherwise stated, all vector and tensor components are components in the coordinate system with base vectors $\boldsymbol{e}_i$. Time is measured from a fixed *reference time* $t=0$. Suppose (see Fig. 4.1) that at $t=0$ a fixed region of space $\mathcal{R}_0$, which may be finite or infinite in extent, is occupied by continuously distributed matter; that is, we suppose that each point of $\mathcal{R}_0$ is occupied by a particle of matter. The material within $\mathcal{R}_0$ at $t=0$ forms a *body* which is denoted by $\mathcal{B}$. Let $\mathbf{X}$ be the position vector, relative to O, of a typical point P_0 within $\mathcal{R}_0$. Then the components X_R of $\mathbf{X}$, in the chosen coordinate system, are the coordinates of the position occupied by a particle of $\mathcal{B}$ at $t=0$. Each point of the region $\mathcal{R}_0$ corresponds to a particle of the body $\mathcal{B}$, and $\mathcal{B}$ is the assemblage of all such particles.

Suppose that the material which occupies the region $\mathcal{R}_0$ at $t=0$ moves so that at a subsequent time t it occupies a new continuous region of space $\mathcal{R}$, and that the material is now continuously distributed in $\mathcal{R}$. This is termed a *motion* of the body $\mathcal{B}$. We make the assumption (which is an essential feature of continuum mechanics) that we can identify individual particles of the body $\mathcal{B}$; that is, we assume that we can identify a point of $\mathcal{R}$ (denoted by P) with position vector $\boldsymbol{x}$, which is occupied at t by the particle which was at P_0 at the time $t=0$. Then the motion of

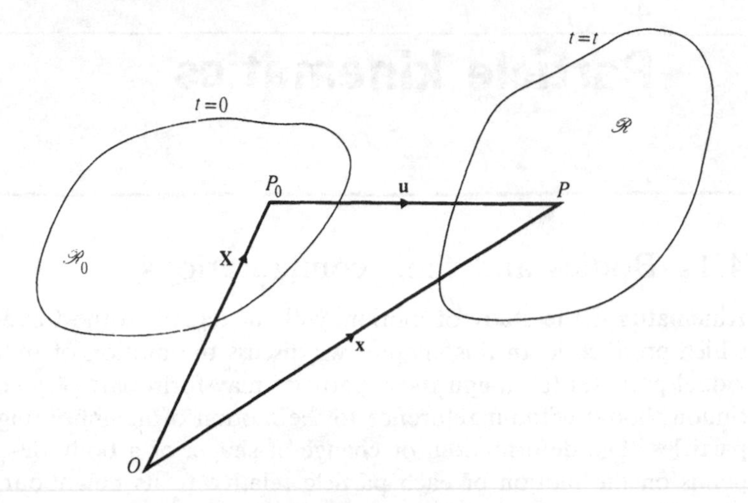

Figure 4.1 Reference and current configurations of the body $\mathscr{B}$

$\mathscr{B}$ can be described by specifying the dependence of the positions $\boldsymbol{x}$ of the particles of $\mathscr{B}$ at time t on their positions $\boldsymbol{X}$ at time $t=0$; that is, by equations of the form

$$\boldsymbol{x}=\boldsymbol{x}(\boldsymbol{X}, t) \tag{4.1}$$

for all $\boldsymbol{X}$ in $\mathscr{R}_0$, and all $\boldsymbol{x}$ in $\mathscr{R}$. If x_i denote the components of $\boldsymbol{x}$ (that is, the coordinates of points in $\mathscr{R}$), then (4.1) may be written in component form as

$$x_i = x_i(X_R, t) \qquad (i, R=1, 2, 3) \tag{4.2}$$

A given particle of the body $\mathscr{B}$ may be distinguished by its coordinates X_R at $t=0$. Thus the coordinates X_R serve as 'labels' with which to identify the particles of $\mathscr{B}$; a particular particle retains the same values of X_R throughout a motion. The coordinates x_i, on the other hand, identify points of space which in general are occupied by different particles at different times. Accordingly, the coordinates X_R are termed *material coordinates* and the coordinates x_i are termed *spatial coordinates.* The set of positions of the particles of $\mathscr{B}$ at a given time specified a *configuration of* $\mathscr{B}$. The configuration of $\mathscr{B}$ at the reference time $t=0$ is its *reference configuration.* Its configuration at time t is its *current configuration* at t.

As far as possible we shall denote scalar, vector and tensor quantities evaluated in the reference configuration by capital letters and corresponding quantities evaluated in the current configuration by lower-case letters. Occasionally we shall employ the index zero (as, for example, in ρ_0) for quantities evaluated in the reference configuration. This convention regarding the use of capital and lower-case letters will extend also to indices of vector and tensor components. Components of vectors and tensors which transform with the coordinates X_R will have capital letter indices (A_R, C_{RS}, etc.) and components which transform with the coordinates x_i will have lower-case indices (a_i, T_{ij}, etc.). Occasionally the convention that capital and lower-case letters relate to the reference and current configurations respectively will conflict with the notation established in Chapter 3; thus in this and subsequent chapters $\mathbf{X}$ is a position vector despite the convention that vectors are normally represented by lower-case italic letters.

For physically realizable motions it is possible in principle to solve (4.2) for X_R in terms of x_i and t, which gives equations of the form

$$\mathbf{X}=\mathbf{X}(\boldsymbol{x}, t), \qquad \text{or} \qquad X_R=X_R(x_i, t) \qquad (R, i=1,2,3) \quad (4.3)$$

Equations (4.3) give the coordinates X_R in the reference configuration of the particle which occupies the position x_i in the current configuration at time t.

Problems in continuum mechanics may be formulated either with the material coordinates X_R as independent variables, in which case we employ the *material description* of the problem, or with the spatial coordinates x_i as independent variables, in which case we employ the *spatial description*. Often the terms 'Lagrangian' and 'Eulerian' are used in place of 'material' and 'spatial' respectively. In the material description attention is focused on what is happening at, or in the neighbourhood of, a particular material particle. In the spatial description we concentrate on events at, or near to, a particular point in space. The mathematical formulation of general physical laws and the description of the properties of particular materials is often most easily accomplished in the material description, but for the solution of particular problems it is frequently preferable to use the spatial description. It is therefore necessary to employ both descriptions, and to relate them to each other. In principle it is possible to transform a problem from the material to the spatial description or vice versa by using (4.2) or (4.3). In practice the transition is not always accomplished easily.

4.2 Displacement and velocity

The displacement vector $\boldsymbol{u}$ of a typical particle from its position $\mathbf{X}$ in the reference configuration to its position $\boldsymbol{x}$ at time t is

$$\boldsymbol{u}=\boldsymbol{x}-\mathbf{X} \tag{4.4}$$

In the material description $\boldsymbol{u}$ is regarded as a function of $\mathbf{X}$ and t, so that

$$\boldsymbol{u}(\mathbf{X}, t)=\boldsymbol{x}(\mathbf{X}, t)-\mathbf{X} \tag{4.5}$$

and in the spatial description $\boldsymbol{u}$ is regarded as a function of $\boldsymbol{x}$ and t, so that

$$\boldsymbol{u}(\boldsymbol{x}, t)=\boldsymbol{x}-\mathbf{X}(\boldsymbol{x}, t) \tag{4.6}$$

The representation (4.5) determines the displacement at time t of the particle defined by the material coordinates X_R. The representation (4.6) determines the displacement which has been undergone by the particle which occupies the position $\boldsymbol{x}$ at time t.

The velocity vector $\boldsymbol{v}$ of a particle is the rate of change of its displacement. Since X_R are constant at a fixed particle it is convenient to employ the material description so that, from (4.5),

$$\boldsymbol{v}(\mathbf{X}, t)=\frac{\partial \boldsymbol{u}(\mathbf{X}, t)}{\partial t}=\frac{\partial \boldsymbol{x}(\mathbf{X}, t)}{\partial t} \tag{4.7}$$

where the differentiations are performed with $\mathbf{X}$ held constant. In terms of the components v_i of $\boldsymbol{v}$, (4.7) may be written as

$$v_i(X_R, t)=\frac{\partial x_i(X_R, t)}{\partial t} \tag{4.8}$$

The result of performing the differentiation (4.7) or (4.8) is to express the velocity components as functions of X_R and t; that is, they give the velocity at time t of the particle which was at $\mathbf{X}$ at $t=0$. We frequently need to employ the spatial description in which we are concerned with the velocity at the point $\boldsymbol{x}$. To do so it is necessary to express v_i in terms of x_i by using the relations (4.3). This is illustrated by the following example:

Example 4.1. A body undergoes the motion defined by

$$x_1=X_1(1+a^2t^2), \qquad x_2=X_2, \qquad x_3=X_3 \tag{4.9}$$

where a is constant. Find the displacement and velocity, in both the material and spatial descriptions.

From (4.5) we have

$$u_1 = X_1 a^2 t^2, \qquad u_2 = 0, \qquad u_3 = 0 \tag{4.10}$$

This gives the displacement at time t in the material description. To obtain the displacement in the spatial description, we substitute for X_1 from (4.9) into (4.10), which gives

$$u_1 = \frac{x_1 a^2 t^2}{1 + a^2 t^2}, \qquad u_2 = 0, \qquad u_3 = 0 \tag{4.11}$$

For the velocity, we differentiate (4.9) with respect to t with X_R fixed to obtain, in the material description,

$$v_1 = 2a^2 X_1 t, \qquad v_2 = 0, \qquad v_3 = 0 \tag{4.12}$$

This is the velocity of the particle which occupied $\boldsymbol{X}$ at $t = 0$. For the spatial description, we eliminate X_1 from (4.9) and (4.12)

$$v_1 = \frac{2a^2 x_1 t}{1 + a^2 t^2}, \qquad v_2 = 0, \qquad v_3 = 0 \tag{4.13}$$

and this gives the velocity of the particle which instantaneously occupies the point $\boldsymbol{x}$ at time t.

4.3 Time rates of change

Suppose that ϕ is some quantity which varies throughout a body in space and in time. We can regard ϕ as a function of t and of either the material coordinates X_R or the spatial coordinates x_i. Thus

$$\phi = G(X_R, t) = g(x_i, t) \tag{4.14}$$

In considering rates of change of ϕ we are usually interested in how ϕ varies with time following a given particle. For example, in Section 4.4 we shall discuss acceleration, which is the rate of change of velocity of a particle. The appropriate quantity to measure the rate of change of ϕ following the particle X_R is $\partial G(X_R, t)/\partial t$, which gives the rate of change of ϕ with X_R held constant. On the other hand, $\partial g(x_i, t)/\partial t$ denotes the rate of change of ϕ with constant x_i (that is, at a fixed point in space) and this is a different quantity.

We adopt the conventional notations $\mathrm{D}\phi/\mathrm{D}t$ or $\dot{\phi}$ for the rate of change of ϕ following a given particle, so that

$$\frac{\mathrm{D}\phi}{\mathrm{D}t} = \dot{\phi} = \frac{\partial G(X_R, t)}{\partial t} \tag{4.15}$$

However, ϕ may be given in the spatial description, so it is necessary to express $\mathrm{D}\phi/\mathrm{D}t$ in terms of derivatives of $g(x_i, t)$. From (4.2) and (4.14) we have

$$\phi = g\{x_i(X_R, t), t\} = g\{x_1(X_R, t), x_2(X_R, t), x_3(X_R, t), t\}$$

Hence, by differentiating with respect to t with X_R constant,

$$\frac{\mathrm{D}\phi}{\mathrm{D}t} = \frac{\partial g(x_i, t)}{\partial x_1}\frac{\partial x_1(X_R, t)}{\partial t} + \frac{\partial g(x_i, t)}{\partial x_2}\frac{\partial x_2(X_R, t)}{\partial t} + \frac{\partial g(x_i, t)}{\partial x_3}\frac{\partial x_3(X_R, t)}{\partial t} + \frac{\partial g(x_i, t)}{\partial t}$$

By using the summation convention, this is written concisely as

$$\frac{\mathrm{D}\phi}{\mathrm{D}t} = \frac{\partial g(x_i, t)}{\partial x_j}\frac{\partial x_j(X_R, t)}{\partial t} + \frac{\partial g(x_i, t)}{\partial t} \tag{4.16}$$

Now by using (4.8), $\mathrm{D}\phi/\mathrm{D}t$ may be written in the simpler form

$$\frac{\mathrm{D}\phi}{\mathrm{D}t} = v_j\frac{\partial g(x_i, t)}{\partial x_j} + \frac{\partial g(x_i, t)}{\partial t} \tag{4.17}$$

or alternatively, in vector notation, as

$$\frac{\mathrm{D}\phi}{\mathrm{D}t} = \boldsymbol{v} \cdot \operatorname{grad} g(x_i, t) + \frac{\partial g(x_i, t)}{\partial t} \tag{4.18}$$

where the gradient is taken with respect to spatial coordinates x_i.

The above is a formal derivation of the formula for $\mathrm{D}\phi/\mathrm{D}t$. To give it a physical interpretation we refer to Fig. 4.2. Consider the change in ϕ following a particle. Suppose that in the time interval t to $t+\delta t$, ϕ (at the particle with coordinates x_i at t) changes its value from ϕ to $\phi+\delta\phi$. During this time interval the particle moves from x_i to $x_i + v_i\,\delta t$, where $\boldsymbol{v}$ is the velocity of the particle at some time between t and $t+\delta t$ (any necessary continuity conditions are assumed to be satisfied). Thus we have to compare the

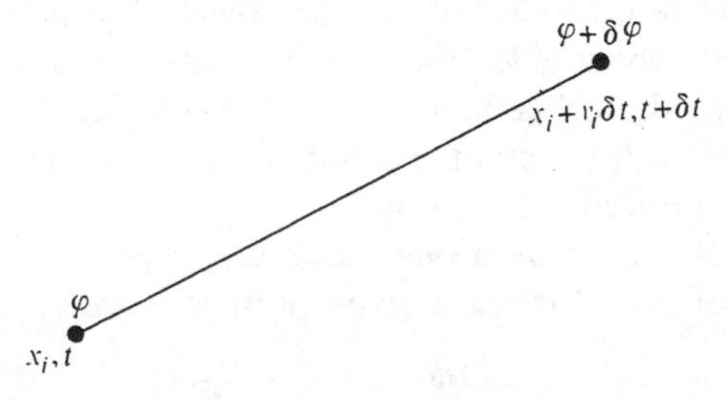

Figure 4.2 The change of ϕ following a particle

value of ϕ at x_i and t, given as $g(x_i, t)$, with its value at $x_i + v_i\,\delta t$ and $t+\delta t$, which is $g(x_i + v_i\,\delta t, t+\delta t)$. Thus

$$\delta\phi = g(x_i + v_i\,\delta t, t+\delta t) - g(x_i, t)$$

Then by applying the mean-value theorem and proceeding to the limit $\delta t \to 0$ in the usual way, it follows that

$$\frac{D\phi}{Dt} = \lim_{\delta t\to 0} \frac{\delta\phi}{\delta t} = v_j \frac{\partial g(x_i, t)}{\delta x_j} + \frac{\partial g(x_i, t)}{\partial t}$$

which is (4.17).

The derivative $D\phi/Dt$ is called the *material derivative* or the *convected derivative* of ϕ.

Although it is logical in (4.14) to use the different symbols G and g for the two functions which describe the dependence of ϕ on the two sets of independent variables (X_R, t) and (x_i, t), it is found in practice that this procedure can lead to a confusing proliferation of symbols. In future we shall adopt the convention of using the same symbol to denote a dependent variable and a function which determines that variable and, where there is a possibility of confusion, the arguments of functions will be explicitly included to demonstrate which independent variables are being employed. Thus, in place of (4.15) we shall write

$$\frac{D\phi}{Dt} = \frac{\partial\phi(X_R, t)}{\partial t} \tag{4.19}$$

and in place of (4.17) and (4.18) we shall write

$$\begin{aligned}\frac{D\phi}{Dt} &= v_j \frac{\partial\phi(x_i, t)}{\partial x_j} + \frac{\partial\phi(x_i, t)}{\partial t}\\ &= \boldsymbol{v}\cdot \operatorname{grad}\phi(x_i, t) + \frac{\partial\phi(x_i, t)}{\partial t}\end{aligned} \tag{4.20}$$

The explicit inclusion of the arguments makes it clear that in (4.19), ϕ is regarded as a function of X_R and t and that in (4.20), ϕ is regarded as a function of x_i and t.

4.4 Acceleration

The acceleration of a particle is the rate of change of velocity of that particle; that is, it is the material derivative of the velocity. We denote the acceleration vector by $\boldsymbol{f}$, and its components by f_i.

Thus, in the material description,

$$x_i = x_i(X_R, t), \qquad v_i = \frac{\partial x_i(X_R, t)}{\partial t},$$

$$f_i = \frac{\partial v_i(X_R, t)}{\partial t} = \frac{\partial^2 x_i(X_R, t)}{\partial t^2} \tag{4.21}$$

or, in vector notation

$$\boldsymbol{v} = \dot{\boldsymbol{x}}(\mathbf{X}, t), \qquad \boldsymbol{f} = \dot{\boldsymbol{v}}(\mathbf{X}, t) = \ddot{\boldsymbol{x}}(\mathbf{X}, t) \tag{4.22}$$

These relations give f in material coordinates. To find the acceleration in terms of spatial coordinates it is necessary to express material coordinates X_R in terms of spatial coordinates x_i. Frequently this information is not explicitly available.

Although (4.21) give the simplest expressions for f_i, they are not the most generally useful, because it is often required to express the acceleration components in terms of derivatives of the velocity components, when the velocity components are expressed in spatial coordinates x_i. Thus, from the results of Section 4.3,

$$f_i = \frac{\mathrm{D}v_i}{\mathrm{D}t} = \frac{\partial v_i(x_j, t)}{\partial t} + v_k \frac{\partial v_i(x_j, t)}{\partial x_k} \tag{4.23}$$

The physical interpretation of this expression is as follows. In an increment of time δt the particle which at time t has coordinates x_k moves to $x_k + v_k\,\delta t$. Hence the velocity components of this particle change from $v_i(x_k, t)$ to $v_i(x_k + v_k\,\delta t, t + \delta t)$. Thus the change in $\boldsymbol{v}$ at a particle is given by

$$\delta v_i = v_i(x_k + v_k\,\delta t, t + \delta t) - v_i(x_k, t)$$

and (4.23) follows by applying the mean-value theorem and proceeding to the limit $\delta t \to 0$. The expression (4.23) gives f_i in terms of the spatial coordinates x_i.

Example 4.2. To illustrate the equivalence of the expressions (4.21) and (4.23) for f_i, consider the motion (4.9). This gives (Example 4.1)

$$v_1 = 2X_1a^2t = \frac{2x_1a^2t}{1 + a^2t^2}, \qquad v_2 = 0, \qquad v_3 = 0$$

By taking the first expression for v_1, we find from (4.21) that

$$f_1 = 2X_1a^2, \qquad f_2 = 0, \qquad f_3 = 0. \tag{4.24}$$

If v_1 is given in the spatial description as $2x_1a^2t/(1+a^2t^2)$, we obtain from (4.23),

$$f_1 = \frac{\partial}{\partial t}\left\{\frac{2x_1a^2t}{1+a^2t^2}\right\} + \frac{2x_1a^2t}{1+a^2t}\frac{\partial}{\partial x_1}\left\{\frac{2x_1a^2t}{1+a^2t^2}\right\}$$

$$= \frac{2x_1a^2(1-a^2t^2)}{(1+a^2t^2)^2} + \frac{2x_1a^2t}{(1+a^2t^2)}\frac{2a^2t}{(1+a^2t^2)} = \frac{2x_1a^2}{1+a^2t^2} \qquad (4.25)$$

$$f_2 = 0, \qquad f_3 = 0$$

The expressions for f_1 given by (4.24) and (4.25) are the same because, from (4.9), $x_1 = X_1(1+a^2t^2)$.

4.5 Steady motion. Particle paths and streamlines

A motion is said to be *steady* if the velocity at any point is independent of time, so that $\boldsymbol{v} = \boldsymbol{v}(\boldsymbol{x})$. Conditions approximating to steady motion are achieved in many practical situations; for example, in flow of a fluid through a pipe at a uniform rate, or flow past a fixed obstacle with uniform velocity at a large distance from the obstacle.

A motion may be unsteady in relation to a fixed coordinate system but steady relative to suitably chosen moving axes. For example, the flow past an aeroplane moving at constant velocity through a uniform atmosphere is unsteady relative to fixed coordinates, but is steady relative to axes which are fixed in relation to the aeroplane and move with it.

The equations (4.2), $x_i = x_i(X_R, t)$, give the successive positions x_i of the particle X_R, with t serving as a parameter. Thus they are parametric equations of the path of the particle X_R. In differential form, (4.2) gives

$$\mathrm{d}x_i = v_i(X_R, t)\,\mathrm{d}t$$

and this can be expressed in spatial coordinates as

$$\mathrm{d}x_i = v_i\{X_R(x_j, t), t\}\,\mathrm{d}t \qquad (4.26)$$

The streamlines at time t are space curves whose tangents are everywhere directed along the direction of the velocity vector. Thus they are given, in terms of a parameter τ, by the equations

$$\mathrm{d}x_i = v_i(x_j, t)\,\mathrm{d}\tau \qquad (4.27)$$

In general, the particle paths and streamlines do not coincide. However, if the motion is steady, so that $\boldsymbol{v}$ is independent of t, then (4.26) and (4.27) represent the same families of curves, and then the particle paths and streamlines are coincident.

4.6 Problems

1. A motion of a fluid is given by the equations

$$x_1 = X_1 + X_2 t + X_3 t^2$$
$$x_2 = X_2 + X_3 t + X_1 t^2$$
$$x_3 = X_3 + X_1 t + X_2 t^2$$

Find the velocity and acceleration of: (a) the particle which was at the point $(1, 1, 1)$ at the reference time $t = 0$, and (b) the particle which occupies the point $(1, 1, 1)$ at time t. Explain why this motion becomes physically unrealistic as $t \to 1$.

2. The velocity in a steady helical flow of a fluid is given by

$$v_1 = -Ux_2, \qquad v_2 = Ux_1, \qquad v_3 = V$$

where U and V are constants. Show that $\operatorname{div} \boldsymbol{v} = 0$ and find the acceleration of the particle at $\boldsymbol{x}$. Also determine the streamlines.

3. The velocity at a point $\boldsymbol{x}$ in space in a body of fluid in steady flow is given by

$$\boldsymbol{v} = U \frac{a^2(x_1^2 - x_2^2)}{(x_1^2 + x_2^2)^2} \boldsymbol{e}_1 + 2U \frac{a^2 x_1 x_2}{(x_1^2 + x_2^2)^2} \boldsymbol{e}_2 + V \boldsymbol{e}_3$$

where U, V and a are constants. Show that $\operatorname{div} \boldsymbol{v} = 0$ and find the acceleration of the particle at $\boldsymbol{x}$. Also determine the streamlines.

4. An electromagnetic fluid is subjected to a decaying electric field of magnitude $\phi = r^{-1} e^{-At}$, where $r^2 = x_1^2 + x_2^2 + x_3^2$ and A is constant. The velocity of the fluid is $\boldsymbol{v} = x_1 x_3 \boldsymbol{e}_1 + x_2^2 t \boldsymbol{e}_2 + x_2 x_3 t \boldsymbol{e}_3$. Determine: (a) the rate of change of ϕ at $t = 1$ of the particle which occupies the point with coordinates $(2, -2, 1)$; (b) the acceleration of the same particle at the same time; (c) the position of the same particle at all subsequent times t. Write down the differential equations of the streamlines and show that at each instant x_2/x_3 is constant along a given streamline.

5. Given the velocity field

$$v_1 = \frac{a_1 x_1 + a_2 x_2}{1+t}, \qquad v_2 = \frac{b_1 x_1 + b_2 x_2}{1+t}, \qquad v_3 = \frac{c x_3}{1+t}$$

with a_1, a_2, b_1, b_2 and c constants, show that the x_2 component of the acceleration at $t=0$ is $(a_1 b_1 + b_1 b_2 - b_1)X_1 + (b_2^2 + b_1 a_2 - b_2)X_2$, where $\mathbf{X}$ denotes the position vector at $t=0$. In the case $a_1 = A$, $a_2 = 0$, $b_1 = 0$, $b_2 = 2A$, $c = 3A$, obtain the particle paths and the streamlines, and show that in this case they coincide.

5

Stress

5.1 Surface traction

In this chapter we consider the forces acting in the interior of a continuous body. Suppose that part of a body $\mathscr{B}$ occupies a region $\mathscr{R}$ which has surface $\mathscr{S}$ as illustrated in Fig. 5.1. Let P be a point on the surface $\mathscr{S}$, $\boldsymbol{n}$ a unit vector directed along the outward normal to $\mathscr{S}$ at P, and δS the area of an element of $\mathscr{S}$ which contains P. We assume that $\mathscr{S}$ and $\mathscr{R}$ possess any necessary smoothness and continuity properties; for example, it is assumed that the normal to $\mathscr{S}$ is uniquely defined at P.

It is also assumed that on the surface element with area δS, the material *outside* $\mathscr{R}$ exerts a force

$$\delta \boldsymbol{p} = \boldsymbol{t}^{(n)}\,\delta S \tag{5.1}$$

on the material *inside* $\mathscr{R}$. The force $\delta\boldsymbol{p}$ is called the *surface force* and $\boldsymbol{t}^{(n)}$ the *mean surface traction* transmitted across the element of area δS from the outside to the inside of $\mathscr{R}$. A similar force, equal in magnitude but opposite in direction to $\delta\boldsymbol{p}$, and a similar surface traction, equal in magnitude but opposite in direction to $\boldsymbol{t}^{(n)}$, is transmitted across the element with area δS from the inside to the outside of $\mathscr{R}$.

Clearly $\boldsymbol{t}^{(n)}$ will depend on the position of P and the direction of $\boldsymbol{n}$. It is further assumed that as $\delta S \to 0$, $\boldsymbol{t}^{(n)}$ tends to a finite limit which is independent of the shape of the element with area δS. Henceforth the symbol $\boldsymbol{t}^{(n)}$ is used to denote the limit

$$\boldsymbol{t}^{(n)} = \lim_{\delta S \to 0} \frac{\delta \boldsymbol{p}}{\delta S} \tag{5.2}$$

and we omit the adjective 'mean' and call $\boldsymbol{t}^{(n)}$ the *surface traction* at the point P on the surface with normal $\boldsymbol{n}$.

The assumptions made above are plausible, but they are of a physical nature and can only be justified to the extent that conclusions based on them agree with observations of what happens

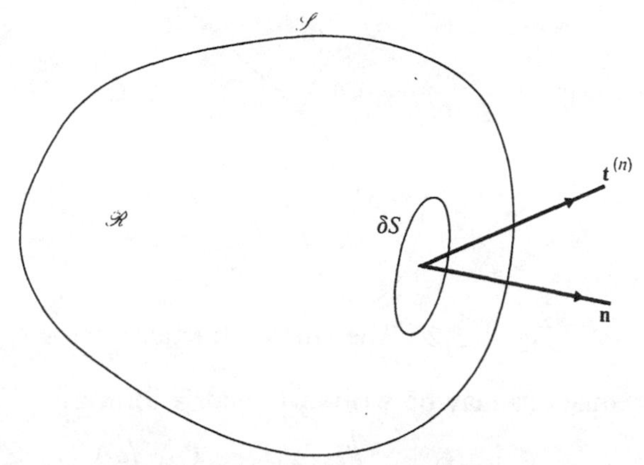

Figure 5.1 Surface traction

to real materials. It is possible for couples as well as forces to be transmitted across a surface. Such couples are of interest but are beyond the scope of this book. In practice their influence is restricted to rather special situations.

It is important to remember that in general $\boldsymbol{t}^{(n)}$ does not coincide in direction with $\boldsymbol{n}$. The force transmitted across a surface does not necessarily act in the direction normal to the surface.

5.2 Components of stress

At P, there is a vector $\boldsymbol{t}^{(n)}$ associated with each direction through P. In particular, given a system of rectangular cartesian coordinates with base vectors $\boldsymbol{e}_i$, there is such a vector associated with the direction of each of the base vectors. Let $\boldsymbol{t}_1$ be the surface traction associated with the direction of $\boldsymbol{e}_1$, from the positive to the negative side (that is, $\boldsymbol{t}_1$ is the force per unit area exerted on the negative side of a surface normal to the x_1-axis by the material on the positive side of this surface; see Fig. 5.2). Surface traction vectors $\boldsymbol{t}_2$ and $\boldsymbol{t}_3$ are similarly defined in relation to the directions of $\boldsymbol{e}_2$ and $\boldsymbol{e}_3$.

Now resolve the vectors $\boldsymbol{t}_1$, $\boldsymbol{t}_2$ and $\boldsymbol{t}_3$ into components in the coordinate system with base vectors $\boldsymbol{e}_i$, as follows:

$$\begin{aligned} \boldsymbol{t}_1 &= T_{11}\boldsymbol{e}_1 + T_{12}\boldsymbol{e}_2 + T_{13}\boldsymbol{e}_3 \\ \boldsymbol{t}_2 &= T_{21}\boldsymbol{e}_1 + T_{22}\boldsymbol{e}_2 + T_{23}\boldsymbol{e}_3 \\ \boldsymbol{t}_3 &= T_{31}\boldsymbol{e}_1 + T_{32}\boldsymbol{e}_2 + T_{33}\boldsymbol{e}_3 \end{aligned} \tag{5.3}$$

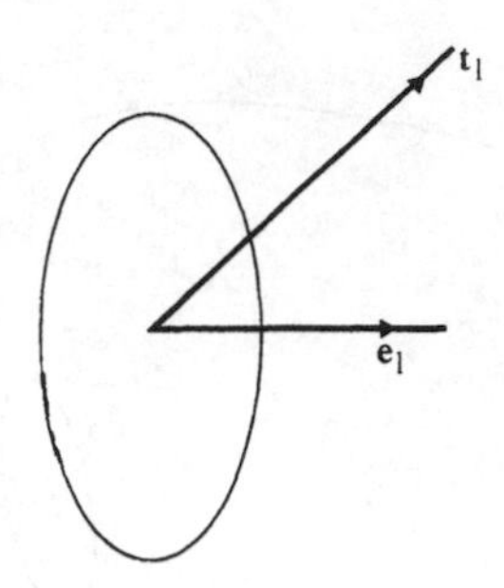

Figure 5.2 The surface traction vector $\boldsymbol{t}_1$

These equations may be written in matrix form as

$$\begin{pmatrix} \boldsymbol{t}_1 \\ \boldsymbol{t}_2 \\ \boldsymbol{t}_3 \end{pmatrix} = \begin{pmatrix} T_{11} & T_{12} & T_{13} \\ T_{21} & T_{22} & T_{23} \\ T_{31} & T_{32} & T_{33} \end{pmatrix} \begin{pmatrix} \boldsymbol{e}_1 \\ \boldsymbol{e}_2 \\ \boldsymbol{e}_3 \end{pmatrix} \tag{5.4}$$

or, using the summation convention, as

$$\boldsymbol{t}_i = T_{ij}\boldsymbol{e}_j \qquad (i, j = 1, 2, 3) \tag{5.5}$$

Since $\boldsymbol{e}_i \cdot \boldsymbol{e}_j = \delta_{ij}$, it follows from (5.5) that

$$T_{ij} = \boldsymbol{t}_i \cdot \boldsymbol{e}_j \tag{5.6}$$

The quantities T_{ij} are called *stress components*. The component T_{11}, for example, is the component of $\boldsymbol{t}_1$ in the direction of $\boldsymbol{e}_1$. T_{11} is positive if the material on the x_1-positive side of the surface on which $\boldsymbol{t}_1$ acts (a surface normal to the x_1-axis) is *pulling* the material on the x_1-negative side. The material is then in *tension* in the x_1 direction. The material on the negative side of the surface is *pulling* in the opposite direction on the material on the positive side. If the material on each side of the surface *pushes* against that on the other, T_{11} is negative, and the material is said to be in *compression* in the x_1 direction. The components T_{11}, T_{22} and T_{33} are called *normal* or *direct stress* components. The remaining components T_{12}, T_{13}, etc., are called *shearing stress* components. All the stress components can be illustrated as the components of forces acting on the faces of a unit cube, as shown in Fig. 5.3.

5.3 The traction on any surface

Suppose that the stress components T_{ij} are known at a given point P. We consider how we may determine the surface traction

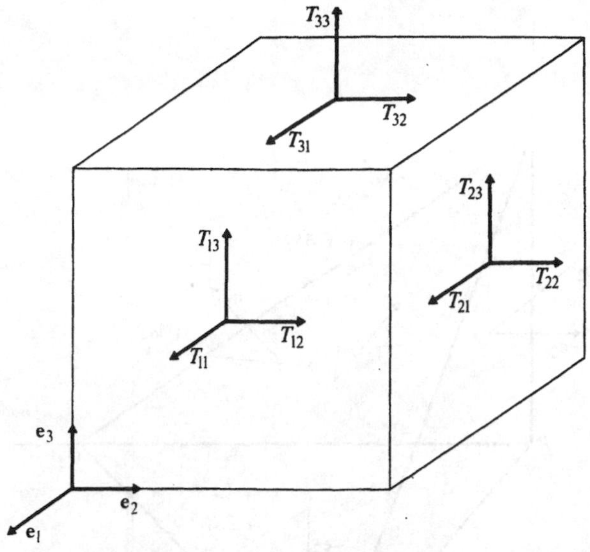

Figure 5.3 Components of the forces on three faces of a unit cube. Opposite forces act on the opposite faces

on an arbitrary surface through P. For this we examine the forces acting on the elementary tetrahedron illustrated in Fig. 5.4. We wish to find the traction $\boldsymbol{t}^{(n)}$ on a surface normal to $\boldsymbol{n}$ at P. In the tetrahedron shown in Fig. 5.4, PQ_1, PQ_2, PQ_3 are parallel to the three coordinate axes, and $Q_1Q_2Q_3$ is normal to $\boldsymbol{n}$. We denote by $-\boldsymbol{t}_1$, $-\boldsymbol{t}_2$, $-\boldsymbol{t}_3$ the mean surface tractions on the faces PQ_2Q_3, PQ_3Q_1 and PQ_1Q_2 respectively. The minus signs arise because we wish to consider the forces acting *on* the tetrahedron, so that, for example, $-\boldsymbol{t}_1$ is the traction exerted on the surface PQ_2Q_3 by material to the left of this surface, on material to the right of the surface; that is, by the material outside the tetrahedron on the material inside the tetrahedron. Similarly, $\boldsymbol{t}^{(n)}$ denotes the mean surface traction on $Q_1Q_2Q_3$ exerted by material on the side towards which $\boldsymbol{n}$ is directed (the outside of the tetrahedron) onto the other side. Let the area of $Q_1Q_2Q_3$ be δS and the volume of $PQ_1Q_2Q_3$ be δV. Then the areas of the other faces are

$$\delta S_1 = n_1\,\delta S, \qquad \delta S_2 = n_2\,\delta S, \qquad \delta S_3 = n_3\,\delta S \tag{5.7}$$

where n_i are the components of $\boldsymbol{n}$; that is, n_i are the directior cosines of the direction of $\boldsymbol{n}$.

The forces exerted on the tetrahedron across its four faces are

$$-\boldsymbol{t}_1\,\delta S_1, \qquad -\boldsymbol{t}_2\,\delta S_2, \qquad -\boldsymbol{t}_3\,\delta S_3, \qquad \boldsymbol{t}^{(n)}\,\delta S$$

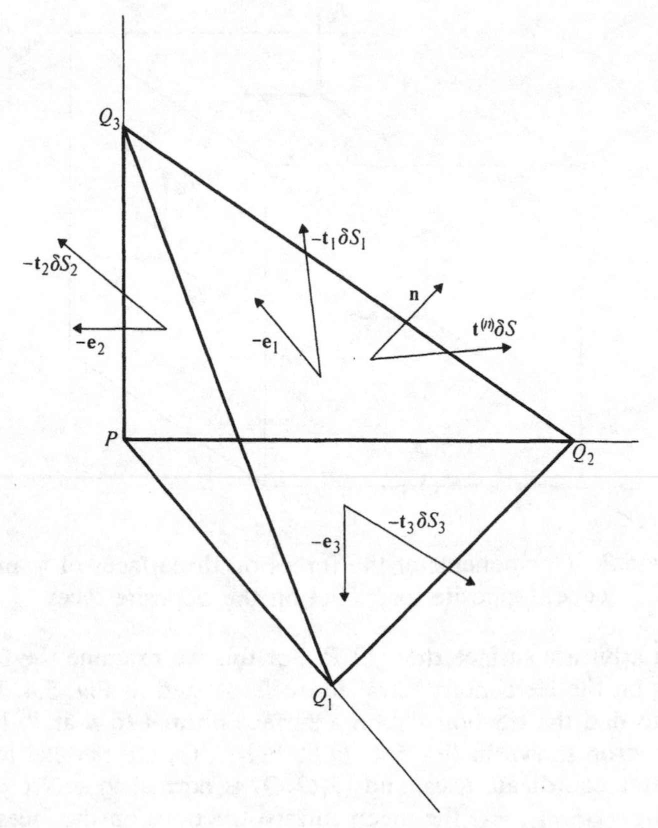

Figure 5.4 Forces acting on an elementary tetrahedron

It is also supposed that there is a *body force* whose mean value over the tetrahedron is $\boldsymbol{b}$ per unit mass, or $\rho\boldsymbol{b}$ per unit volume, where ρ is the density. The most common example of a body force is a gravitational force, but there are other possibilities.

We now assume that for any part of a body, and in particular for the elementary tetrahedron $PQ_1Q_2Q_3$, the rate of change of momentum is proportional to the resultant force acting. Although this is a natural assumption to make, it is a new assumption which is stronger than Newton's second law, for Newton's law applies only to bodies as a whole. Moreover, it is an assumption which cannot be verified directly by experiment, for it is impossible to make direct measurements of internal surface tractions; their existence and magnitudes can only be inferred from observations of other quantities. Nevertheless, the consequences of this assumption (which is sometimes called Cauchy's law of motion) are so well verified that it is hardly open to question.

For the tetrahedron $PQ_1Q_2Q_3$, Cauchy's law gives

$$-\boldsymbol{t}_1\,\delta S_1 - \boldsymbol{t}_2\,\delta S_2 - \boldsymbol{t}_3\,\delta S_3 + \boldsymbol{t}^{(n)}\,\delta S + \rho\boldsymbol{b}\,\delta V = \rho\boldsymbol{f}\,\delta V$$

With (5.7), this may be written as

$$\boldsymbol{t}^{(n)} = \boldsymbol{t}_1 n_1 + \boldsymbol{t}_2 n_2 + \boldsymbol{t}_3 n_3 + \rho\frac{\delta V}{\delta S}(\boldsymbol{f} - \boldsymbol{b})$$

Now, with $\boldsymbol{n}$ and the point P fixed, let the tetrahedron shrink in size but retain its shape. Thus $\delta S \to 0$, and in this limit all quantities are evaluated at P, so that $\boldsymbol{t}_1$, $\boldsymbol{t}_2$, $\boldsymbol{t}_3$ and $\boldsymbol{t}^{(n)}$ become tractions at P, and ρ, $\boldsymbol{b}$ and $\boldsymbol{f}$ are evaluated at P. Also, since δV is proportional to the cube, and δS is proportional to the square, of the linear dimensions of the tetrahedron, $\delta V/\delta S \to 0$ as $\delta S \to 0$. Thus in this limit,

$$\boldsymbol{t}^{(n)} = \boldsymbol{t}_1 n_1 + \boldsymbol{t}_2 n_2 + \boldsymbol{t}_3 n_3 = n_i \boldsymbol{t}_i = n_i T_{ij} \boldsymbol{e}_j \tag{5.8}$$

where the last relation makes use of (5.5). This gives the traction on any surface with unit normal $\boldsymbol{n}$ in terms of the stress components T_{ij}. The components $t_j^{(n)}$ of $\boldsymbol{t}^{(n)}$ are given by

$$t_j^{(n)} = n_i T_{ij} \tag{5.9}$$

The easiest way to calculate $\boldsymbol{t}^{(n)}$ is to use (5.9) in the matrix form

$$\begin{pmatrix} t_1^{(n)} \\ t_2^{(n)} \\ t_3^{(n)} \end{pmatrix} = \begin{pmatrix} T_{11} & T_{21} & T_{31} \\ T_{12} & T_{22} & T_{32} \\ T_{13} & T_{23} & T_{33} \end{pmatrix} \begin{pmatrix} n_1 \\ n_2 \\ n_3 \end{pmatrix} \tag{5.10}$$

A numerical example is given in Example 5.1 in Section 5.6.

5.4 Transformation of stress components

The stress components T_{ij} were defined in Section 5.2 in relation to the coordinate system with base vectors $\boldsymbol{e}_i$. The choice of a different coordinate system will lead to a different set of stress components. We now examine the relationship between the stress components T_{ij} associated with base vectors $\boldsymbol{e}_i$, and stress components $\bar{T}_{ij}$ at the same point but referred to a new coordinate system with base vectors $\bar{\boldsymbol{e}}_i$, where

$$\bar{\boldsymbol{e}}_i = M_{ij}\boldsymbol{e}_j, \qquad \boldsymbol{e}_j = M_{ij}\bar{\boldsymbol{e}}_i \tag{5.11}$$

and $\mathbf{M} = (M_{ij})$ is an orthogonal matrix.

In (5.8), we may as a special case choose $\boldsymbol{n}$ to be $\bar{\boldsymbol{e}}_1$. From (5.11), the components of $\bar{\boldsymbol{e}}_1$ referred to base vectors $\boldsymbol{e}_i$ are M_{11}, M_{12} and M_{13}. We denote by $\bar{\boldsymbol{t}}_1$ the traction on a surface normal to $\bar{\boldsymbol{e}}_1$. Then, from (5.8) (with $n_i = M_{1i}$) and (5.11),

$$\bar{\boldsymbol{t}}_1 = M_{1i}\boldsymbol{t}_i = M_{1i}T_{ij}\boldsymbol{e}_j = M_{1i}M_{qj}T_{ij}\bar{\boldsymbol{e}}_q$$

We define $\bar{\boldsymbol{t}}_2$ and $\bar{\boldsymbol{t}}_3$ in a similar way, and obtain similar relations for them. The general relation is

$$\bar{\boldsymbol{t}}_p = M_{pi}M_{qj}T_{ij}\bar{\boldsymbol{e}}_q \tag{5.12}$$

However, the stress components $\bar{T}_{pq}$ referred to base vectors $\boldsymbol{e}_q$ are defined by the relation analogous to (5.5) as the components of $\bar{\boldsymbol{t}}_p$, referred to base vectors $\bar{\boldsymbol{e}}_q$, so that

$$\bar{\boldsymbol{t}}_p = \bar{T}_{pq}\bar{\boldsymbol{e}}_q \tag{5.13}$$

Hence, by comparing (5.12) and (5.13),

$$\bar{T}_{pq} = M_{pi}M_{qj}T_{ij} \tag{5.14}$$

This is just the transformation law (3.28) for the components of a second-order tensor. Hence there exists a second-order tensor $\boldsymbol{T} = T_{ij}\,\boldsymbol{e}_i \otimes \boldsymbol{e}_j$ whose components are T_{ij} in the coordinate system with base vectors $\boldsymbol{e}_i$ and $\bar{T}_{ij}$ in the system with base vectors $\bar{\boldsymbol{e}}_i$. $\boldsymbol{T}$ is called the *Cauchy stress tensor*, and it completely describes the state of stress of a body. Some other stress tensors will be considered briefly in Section 9.5, but we shall not use them in this book, and so we shall refer to $\boldsymbol{T}$ as simply the *stress tensor.*

Equation (5.14) is an important result, because it shows that T_{ij} are components of a tensor, so we briefly recapitulate the steps which lead to (5.14). They are:

(a) define T_{ij} by (5.3), using base vectors $\boldsymbol{e}_i$;
(b) derive the expression (5.8) for the traction on a surface with normal $\boldsymbol{n}$;
(c) take $\boldsymbol{n}$ to be the new base vectors $\bar{\boldsymbol{e}}_1$, $\bar{\boldsymbol{e}}_2$, $\bar{\boldsymbol{e}}_3$ in turn, and so obtain (5.12);
(d) resolve the traction on the new coordinate surfaces in the directions of the new coordinate axes to define $\bar{T}_{pq}$ as in (5.13), and compare with (5.12).

If $\mathbf{T} = (T_{ij})$, and $\bar{\mathbf{T}} = (\bar{T}_{ij})$, then (5.14) may be written in matrix notation as

$$\bar{\mathbf{T}} = \mathbf{M}\mathbf{T}\mathbf{M}^{\mathrm{T}} \tag{5.15}$$

Thus the calculation of stress components in a new coordinate system can be carried out by matrix multiplications, and this is usually the most convenient way to perform such calculations. A numerical example is given in Problem 1 in Section 5.10.

Since it is now established that T_{ij} are components of a tensor, equation (5.9) can be expressed in direct notation as

$$\boldsymbol{t}^{(n)} = \boldsymbol{n} \cdot \boldsymbol{T} \tag{5.16}$$

5.5 Equations of equilibrium

We now consider that the body $\mathscr{B}$ is in equilibrium. The notation of Section 5.1 is used (see Fig. 5.1). $\mathscr{R}$ is an arbitrary region in $\mathscr{B}$ and $\mathscr{S}$ is the surface of $\mathscr{R}$, with unit normal $\boldsymbol{n}$. We assume that in equilibrium the resultant force, and the resultant couple about O, acting on the material in $\mathscr{R}$ are zero. The forces acting on the material in $\mathscr{R}$ are of two kinds: there are the surface forces acting across $\mathscr{S}$ whose resultant is the integral of $\boldsymbol{t}^{(n)}$ over $\mathscr{S}$, and body forces $\rho\boldsymbol{b}$ per unit volume whose resultant is the integral of $\rho\boldsymbol{b}$ through $\mathscr{R}$. Thus the condition for the resultant force to be zero is

$$\iint_{\mathscr{S}} \boldsymbol{t}^{(n)}\, \mathrm{d}s + \iiint_{\mathscr{R}} \rho\boldsymbol{b}\, \mathrm{d}V = \boldsymbol{0} \tag{5.17}$$

Similarly, the resultant couple about O is zero if

$$\iint_{\mathscr{S}} \boldsymbol{x} \times \boldsymbol{t}^{(n)}\, \mathrm{d}S + \iiint_{\mathscr{R}} \rho\boldsymbol{x} \times \boldsymbol{b}\, \mathrm{d}V = \boldsymbol{0} \tag{5.18}$$

where $\boldsymbol{x}$ denotes the position vector relative to O.

In terms of components, (5.17) and (5.18) may be written (with the aid of (5.9)) as

$$\iint_{\mathscr{S}} n_i T_{ij}\, \mathrm{d}S + \iiint_{\mathscr{R}} \rho b_j\, \mathrm{d}V = 0 \tag{5.19}$$

$$\iint_{\mathscr{S}} e_{jpq} x_p n_r T_{rq}\, \mathrm{d}S + \iiint_{\mathscr{R}} \rho e_{jpq} x_p b_q\, \mathrm{d}V = 0 \tag{5.20}$$

We next transform the surface integrals into volume integrals by use of the divergence theorem (Section 3.10). It is assumed that T_{ij} have continuous first derivatives. Then (5.19) and (5.20) become

$$\iiint_{\mathscr{R}} \left(\frac{\partial T_{ij}}{\partial x_i} + \rho b_j\right) \mathrm{d}V = 0 \tag{5.21}$$

$$\iiint_{\mathscr{R}} e_{jpq}\left\{\frac{\partial}{\partial x_r}(x_p T_{rq}) + \rho x_p b_q\right\} \mathrm{d}V = 0 \tag{5.22}$$

However, these relations must hold in every region $\mathscr{R}$ which lies in $\mathscr{B}$. Hence the integrands must be zero throughout $\mathscr{B}$, for if they were not, it would be possible to find a region $\mathscr{R}$ for which (5.21) or (5.22) was violated. Hence, throughout $\mathscr{B}$,

$$\frac{\partial T_{ij}}{\partial x_i} + \rho b_j = 0 \tag{5.23}$$

$$e_{jpq}\left\{\frac{\partial}{\partial x_r}(x_p T_{rq}) + \rho x_p b_q\right\} = 0 \tag{5.24}$$

However, $\partial x_p/\partial x_r = \delta_{pr}$, and so (5.24) may be written as

$$e_{jpq}\left\{x_p\left(\frac{\partial T_{rq}}{\partial x_r} + \rho b_q\right) + T_{pq}\right\} = 0$$

and, by using (5.23), this reduces to

$$e_{jpq} T_{pq} = 0$$

which implies that

$$T_{pq} = T_{qp} \tag{5.25}$$

Equation (5.23) is the *equation of equilibrium*. Equations (5.25) show that in equilibrium the *stress tensor is a symmetric tensor.* In Section 7.5 it will be shown that (5.25) also holds for a body in motion; we anticipate this result and henceforth treat $\boldsymbol{T}$ as a symmetric tensor. Equation (5.23) is given in full in (5.37).

5.6 Principal stress components, principal axes of stress and stress invariants

In general, the surface traction $\boldsymbol{t}^{(n)}$ associated with a direction $\boldsymbol{n}$ through a point P will not act in the direction of the vector $\boldsymbol{n}$; the traction will have a tangential (shearing) component on the surface normal to $\boldsymbol{n}$ as well as a normal component. However, it may happen that for certain special directions $\boldsymbol{n}$ the traction $\boldsymbol{t}^{(n)}$ does act in the direction $\boldsymbol{n}$. We investigate this possibility.

If $\boldsymbol{t}^{(n)}$ and $\boldsymbol{n}$ have the same direction, then

$$\boldsymbol{t}^{(n)} = T\boldsymbol{n}$$

where T is the magnitude of $\boldsymbol{t}^{(n)}$. From (5.16), remembering that $\boldsymbol{T}$ is symmetric, this may be written as

$$\boldsymbol{n} \cdot \boldsymbol{T} = T\boldsymbol{n}$$

or, in components, as

$$n_i T_{ij} = T n_j$$

that is,

$$(T_{ij} - T\delta_{ij}) n_i = 0$$

Hence (Section 3.8), T is one of the three principal components T_1, T_2 and T_3 of $\boldsymbol{T}$, and $\boldsymbol{n}$ determines the corresponding principal axis. Let the unit vectors in the directions of the principal axes be $\boldsymbol{n}_1$, $\boldsymbol{n}_2$ and $\boldsymbol{n}_3$. If these three orthogonal vectors are taken as base vectors at P then, referred to these axes, the matrix of the stress components is a diagonal matrix with diagonal elements T_1, T_2 and T_3. The principal components are the roots of the equation

$$\det(T_{ij} - T\delta_{ij}) = 0 \tag{5.26}$$

where T_{ij} are the stress components referred to any coordinate system. In general, the principal directions vary from point to point, so that it is not usually possible to find a rectangular cartesian coordinate system in which the matrix of stress components is a diagonal matrix everywhere.

Let T_1, T_2 and T_3 be ordered so that $T_1 \geqslant T_2 \geqslant T_3$. It is shown in Example 5.2 that, as the orientation of a surface through P varies, T_1 is the greatest, and T_3 is the least, normal component of the traction on the surface. This property can be used to give an alternative definition of the principal stress components and principal axes of stress.

If (5.26) has two or three equal roots, the above statements remain true, but the principal axes are not uniquely defined.

Example 5.1. The components of the stress tensor at a point P are given in appropriate units by

$$\mathbf{T} = \begin{pmatrix} 1 & 2 & 3 \\ 2 & 4 & 6 \\ 3 & 6 & 1 \end{pmatrix}$$

Find:

(i) the traction $\boldsymbol{t}$ at P on the plane normal to the x_1-axis;
(ii) the traction $\boldsymbol{t}$ at P on the plane whose normal has direction ratios $1:-1:2$;
(iii) the traction $\boldsymbol{t}$ at P on the plane through P parallel to the plane $2x_1-2x_2-x_3=0$;
(iv) the normal component of the traction on the plane (iii);
(v) the principal stress components at P;
(vi) the directions of the principal axes of stress through P.

(i) The plane normal to the x_1-axis has unit normal $(1, 0, 0)$. Hence the traction components on this plane are given by (5.10) as

$$\mathbf{t}=\begin{pmatrix}1&2&3\\2&4&6\\3&6&1\end{pmatrix}\begin{pmatrix}1\\0\\0\end{pmatrix}=\begin{pmatrix}1\\2\\3\end{pmatrix}$$

(ii) The unit normal is $(1, -1, 2)/\sqrt{6}$. Hence

$$\mathbf{t}=\frac{1}{\sqrt{6}}\begin{pmatrix}1&2&3\\2&4&6\\3&6&1\end{pmatrix}\begin{pmatrix}1\\-1\\2\end{pmatrix}=\frac{1}{\sqrt{6}}\begin{pmatrix}5\\10\\-1\end{pmatrix}$$

(iii) The unit normal is $\frac{1}{3}(2, -2, -1)$. Hence

$$\mathbf{t}=\frac{1}{3}\begin{pmatrix}1&2&3\\2&4&6\\3&6&1\end{pmatrix}\begin{pmatrix}2\\-2\\-1\end{pmatrix}=\frac{1}{3}\begin{pmatrix}-5\\-10\\-7\end{pmatrix}$$

(iv) The required component is $\mathbf{n}\cdot\mathbf{t}=\frac{1}{9}\{2\times(-5)-2\times(-10)-1\times(-7)\}=\frac{17}{9}$.

(v) The principal components are solutions of

$$\begin{vmatrix}1-T&2&3\\2&4-T&6\\3&6&1-T\end{vmatrix}=0$$

which gives $T_1=10$, $T_2=0$, $T_3=-4$.

(vi) The principal direction corresponding to, for example, $T_1=10$ is given by the solution of

$$-9n_1+2n_2+3n_3=0$$

$$2n_1+-6n_2+6n_3=0$$

$$3n_1+6n_2+-9n_3=0$$

which give the direction ratios $n_1:n_2:n_3=3:6:5$. Similarly, the direction ratios of the other two principal directions are $-2:1:0$ and $1:2:-3$ (note that these directions are mutually orthogonal).

Example 5.2. Prove that as the orientation of a surface through P varies, T_1 is the greatest, and T_3 is the least, normal component of traction on the surface (assume that T_1, T_2 and T_3 are all different).

Choose the coordinate axes to coincide with the principal axes of $\boldsymbol{T}$, so that the matrix of stress components takes the form

$$\mathbf{T}=\begin{pmatrix} T_1 & 0 & 0 \\ 0 & T_2 & 0 \\ 0 & 0 & T_3 \end{pmatrix}$$

The normal component of traction on a surface with unit normal $\boldsymbol{n}$ is $T_{ij}n_in_j$, which, when $\mathbf{T}$ has the given diagonal form, reduces to $T=T_1n_1^2+T_2n_2^2+T_3n_3^2$. Hence we require extremal values of T for variations of n_1, n_2 and n_3, subject to the constraint $n_1^2+n_2^2+n_3^2=1$. These extrema are given by

$$\begin{aligned} T_1n_1-\sigma n_1&=0\\ T_2n_2-\sigma n_2&=0\\ T_3n_3-\sigma n_3&=0\\ n_1^2+n_2^2+n_3^2&=1 \end{aligned}$$

where σ is a Lagrangian multiplier. The solutions of these equations are

(i) $\mathbf{n}=(\pm 1 \quad 0 \quad 0)^{\mathrm{T}}$, $T=T_1$;
(ii) $\mathbf{n}=(0 \quad \pm 1 \quad 0)^{\mathrm{T}}$, $T=T_2$;
(iii) $\mathbf{n}=(0 \quad 0 \quad \pm 1)^{\mathrm{T}}$, $T=T_3$.

Since $T_1>T_2>T_3$, (i) gives the maximum and (iii) gives the minimum values of T.

As $\boldsymbol{T}$ is a symmetric second-order tensor, the discussion of Section 3.8 shows that $\boldsymbol{T}$ has three independent invariants. We denote these by J_1, J_2, and J_3, where

$$\begin{aligned} J_1&=T_1+T_2+T_3=\mathrm{tr}\,\boldsymbol{T}=T_{ii}\\ J_2&=-(T_2T_3+T_3T_1+T_1T_2)=\tfrac{1}{2}\{\mathrm{tr}\,\boldsymbol{T}^2-(\mathrm{tr}\,\boldsymbol{T})^2\}=\tfrac{1}{2}(T_{ij}T_{ij}-T_{ii}T_{jj}) \qquad (5.27)\\ J_3&=T_1T_2T_3=\det\boldsymbol{T} \end{aligned}$$

Note that the definition of J_2 is not quite consistent with that of I_2 in (3.57), because there is a difference of sign, which it is found convenient to introduce.

5.7 The stress deviator tensor

It is often useful to decompose $\boldsymbol{T}$ in the following way:

$$\boldsymbol{T} = \boldsymbol{S} + \tfrac{1}{3}J_1\boldsymbol{I} \tag{5.28}$$

where $\boldsymbol{S}$ is the *stress deviator tensor.* If S_{ij} denote the components of $\boldsymbol{S}$, then

$$T_{ij} = -p\delta_{ij} + S_{ij} \tag{5.29}$$

where

$$p = -\tfrac{1}{3}T_{kk} = -\tfrac{1}{3}J_1 = -\tfrac{1}{3}\,\mathrm{tr}\,\boldsymbol{T} \tag{5.30}$$

and hence

$$S_{ij} = T_{ij} - \tfrac{1}{3}T_{kk}\delta_{ij} \tag{5.31}$$

and

$$S_{ii} = 0 \tag{5.32}$$

If $S_{ij} = 0$, then the stress has the form $T_{ij} = -p\delta_{ij}$. This is called a *pure hydrostatic* state of stress, and p is the *hydrostatic pressure.* The negative sign arises because we conventionally regard pressure, which causes compression, as positive, but we define compressive stress as negative.

The principal axes of $\boldsymbol{S}$ are the same as those of $\boldsymbol{T}$. If the principal components of $\boldsymbol{S}$ are S_1, S_2, S_3, then

$$S_1 + S_2 + S_3 = 0 \tag{5.33}$$

and

$$S_1 = \tfrac{1}{3}(2T_1 - T_2 - T_3), \qquad S_2 = \tfrac{1}{3}(2T_2 - T_3 - T_1), \qquad S_3 = \tfrac{1}{3}(2T_3 - T_1 - T_2) \tag{5.34}$$

Because S_1, S_2 and S_3 satisfy (5.33), there are only two basic invariants of $\boldsymbol{S}$. These are taken to be J_2' and J_3', where

$$\begin{aligned} J_2' &= -(S_2S_3 + S_3S_1 + S_1S_2) = \tfrac{1}{2}\,\mathrm{tr}\,\boldsymbol{S}^2 \\ J_3' &= S_1S_2S_3 = \det \boldsymbol{S} = \tfrac{1}{3}\,\mathrm{tr}\,\boldsymbol{S}^3 \end{aligned} \tag{5.35}$$

The invariants J_2' and J_3' can be expressed in terms of J_1, J_2 and J_3 by, in (3.63), replacing I_1, I_2, I_3, I_2' and I_3' by J_1, $-J_2$, J_3, $-J_2'$ and J_3' respectively. It is sometimes convenient to adopt J_1, J_2' and J_3' as a set of basic invariants of $\boldsymbol{T}$.

5.8 Shear stress

The normal stress component on a surface normal to the x_1-axis is T_{11} (see Fig. 5.3). The shear stress on this surface is the resultant of the other two components $T_{12}\boldsymbol{e}_2$ and $T_{13}\boldsymbol{e}_3$ of the traction on the surface. Hence the shear stress has magnitude $(T_{12}^2+T_{13}^2)^{\frac{1}{2}}$, and acts in a direction which lies in the surface.

For a general surface with unit normal vector $\boldsymbol{n}$, the normal component of the traction $\boldsymbol{t}^{(n)}$ has magnitude $\boldsymbol{n}\cdot\boldsymbol{t}^{(n)}=n_in_jT_{ij}$. The shear stress on this surface is the component of $\boldsymbol{t}^{(n)}$ normal to $\boldsymbol{n}$, namely

$$\boldsymbol{t}^{(n)}-(\boldsymbol{n}\cdot\boldsymbol{t}^{(n)})\boldsymbol{n}=T_{rs}n_r(\delta_{sj}-n_sn_j)\boldsymbol{e}_j$$

Suppose that the principal stress components are ordered so that $T_1\geqslant T_2\geqslant T_3$, and let the corresponding unit vectors in the directions of the principal axes be $\boldsymbol{n}_1$, $\boldsymbol{n}_2$ and $\boldsymbol{n}_3$. Then it can be shown that, as $\boldsymbol{n}$ varies at point P, the magnitude of the shear stress on the surface normal to $\boldsymbol{n}$ reaches a maximum value $\frac{1}{2}(T_1-T_3)$ when $\boldsymbol{n}$ lies along either of the bisectors of the angle between $\boldsymbol{n}_1$ and $\boldsymbol{n}_3$. The proof resembles that of Example 5.2 and is left as an exercise (Problem 9). Note that $\frac{1}{2}(T_1-T_3)=\frac{1}{2}(S_1-S_3)$, and that in a hydrostatic state of stress $T_1=T_2=T_3$, and then the shear stress is zero on any surface.

5.9 Some simple states of stress

(a) *Hydrostatic pressure.* Suppose that

$$T_{ij}=-p\delta_{ij}$$

that is,

$$T_{11}=T_{22}=T_{33}=-p, \qquad T_{23}=T_{31}=T_{12}=0 \tag{5.36}$$

Then we have a state of hydrostatic pressure. The stress components take the form (5.36) in any rectangular cartesian coordinate system, and any three mutually orthogonal directions may be regarded as principal directions. This is the state of stress in any fluid in equilibrium (that is, in hydrostatics), or in an inviscid fluid whether it is in equilibrium or not. The pressure p is, in general, a function of position.

In the remaining examples, body forces will be regarded as negligible and we seek stress states which satisfy the equilibrium equations (5.23), which are

$$\left.\begin{aligned} \frac{\partial T_{11}}{\partial x_1}+\frac{\partial T_{21}}{\partial x_2}+\frac{\partial T_{31}}{\partial x_3}+\rho b_1=0 \\ \frac{\partial T_{12}}{\partial x_1}+\frac{\partial T_{22}}{\partial x_2}+\frac{\partial T_{32}}{\partial x_3}+\rho b_2=0 \\ \frac{\partial T_{13}}{\partial x_1}+\frac{\partial T_{23}}{\partial x_2}+\frac{\partial T_{33}}{\partial x_3}+\rho b_3=0 \end{aligned}\right\} \tag{5.37}$$

Since these are three equations for the six components of stress, they are insufficient to determine the solution to any problem. Nevertheless, they must be satisfied for any body in equilibrium, and it is of interest to examine some stress states which satisfy them. When the body force is neglected, they are satisfied if the T_{ij} are all constants, in which case the stress is *homogeneous*. The next two examples are in this category.

(b) *Uniform tension or compression* in the x_1 direction is given by

$$T_{11}=\sigma, \qquad T_{22}=T_{33}=T_{23}=T_{31}=T_{12}=0 \tag{5.38}$$

where σ is constant. This gives the stress in a uniform cylindrical bar with generators parallel to the x_1-axis, no forces applied to its lateral surfaces, and uniform forces σ per unit area applied to plane ends normal to the generators. If σ is positive, the bar is in *tension*, and if σ is negative, the bar is in *compression*. The principal stress directions are the x_1 direction and any two directions orthogonal to each other and to the x_1 direction.

(c) *Uniform shear stress* in the x_1 direction on planes $x_2=$constant arises if

$$T_{21}=\tau, \qquad T_{11}=T_{22}=T_{33}=T_{23}=T_{31}=0 \tag{5.39}$$

where τ is constant. This may occur, for example, in laminar shear flow of a viscous fluid, when the fluid flows in the x_1 direction by shearing on the planes $x_2=$constant. The principal axes of stress have the directions of the x_3-axis and the two bisectors of the x_1- and x_2-axes.

(d) *Pure bending.* Let

$$T_{11}=cx_2, \qquad T_{22}=T_{33}=T_{23}=T_{31}=T_{12}=0 \tag{5.40}$$

where c is constant. This approximates the stress in a prismatic beam, with generators parallel to the x_1-axis, which is bent by end couples applied to its ends and acting about axes parallel to the x_3-axis. The plane $x_2=0$ is chosen so that the resultant force on each end is zero. If $c>0$ the region $x_2>0$ of the beam is in tension, and the region $x_2<0$ is in compression. The principal stress directions are as in (b) above.

(e) *Plane stress.* If

$$T_{33}=T_{13}=T_{23}=0 \tag{5.41}$$

and T_{11}, T_{22} and T_{12} are functions only of x_1 and x_2, we have a state of plane stress. In the absence of body forces, the equilibrium equations reduce to

$$\frac{\partial T_{11}}{\partial x_1}+\frac{\partial T_{21}}{\partial x_2}=0, \qquad \frac{\partial T_{12}}{\partial x_1}+\frac{\partial T_{22}}{\partial x_2}=0 \tag{5.42}$$

This is the approximate state of stress in a thin flat plate lying parallel to the x_3-plane, and subject to forces acting in its plane. The x_3 direction is a principal direction; the other two principal directions are in the plane of the plate.

(f) *Pure torsion.* Suppose that

$$T_{13}=x_2g(r), \qquad T_{23}=-x_1g(r), \qquad T_{11}=T_{22}=T_{33}=T_{12}=0 \tag{5.43}$$

where $r^2=x_1^2+x_2^2$. This corresponds to the state of stress in a circular cylindrical bar whose axis coincides with the x_3-axis and which is twisted by couples acting about the axis of the cylinder and applied to the ends of the cylinder, with no forces acting on the curved surfaces. The principal directions are the radial direction and the bisectors of the tangential and axial directions.

5.10 Problems

1. The components of the stress tensor in a rectangular cartesian coordinate system x_1, x_2, x_3 at a point P are given in appropriate units by

$$(T_{ij}) = \begin{pmatrix} 3 & 2 & 2 \\ 2 & 4 & 0 \\ 2 & 0 & 2 \end{pmatrix}$$

Find: (a) the traction at P on the plane normal to the x_1-axis; (b) the traction at P on the plane whose normal has direction ratios $1:-3:2$; (c) the traction at P on a plane through P parallel to the plane $x_1+2x_2+3x_3=1$; (d) the principal stress components at P; (e) the directions of the principal axes of stress at P. Verify that the principal axes of stress are mutually orthogonal.

The coordinates $\bar{x}_1$, $\bar{x}_2$, $\bar{x}_3$ are related to x_1, x_2, x_3 by

$$\bar{x}_1 = \tfrac{1}{3}(x_1 - 2x_2 + 2x_3), \qquad \bar{x}_2 = \tfrac{1}{3}(-2x_1 + x_2 + 2x_3),$$
$$\bar{x}_3 = \tfrac{1}{3}(-2x_1 - 2x_2 - x_3)$$

Verify that this transformation is orthogonal, and find the components of the stress tensor defined above in the new coordinate system. Use the answer to check the answers to (d) and (e) above.

2. In plane stress ($T_{13} = T_{23} = T_{33} = 0$) show that if the $\bar{x}_1$- and $\bar{x}_2$-axes are obtained by rotating the x_1- and x_2-axes through an angle α about the x_3-axis, then

$$\bar{T}_{11} = \tfrac{1}{2}(T_{11} + T_{22}) + \tfrac{1}{2}(T_{11} - T_{22}) \cos 2\alpha + T_{12} \sin 2\alpha$$
$$\bar{T}_{22} = \tfrac{1}{2}(T_{11} + T_{22}) - \tfrac{1}{2}(T_{11} - T_{22}) \cos 2\alpha - T_{12} \sin 2\alpha$$
$$\bar{T}_{12} = -\tfrac{1}{2}(T_{11} - T_{22}) \sin 2\alpha + T_{12} \cos 2\alpha$$

3. If, in appropriate units

$$(T_{ij}) = \begin{pmatrix} 1 & 0 & 2 \\ 0 & 1 & 0 \\ 2 & 0 & -2 \end{pmatrix}$$

find the principal components of stress, and show that the principal directions which correspond to the greatest and least principal components are both perpendicular to the x_2-axis.

4. A cantilever beam with rectangular cross-section occupies the

region $-a \leqslant x_1 \leqslant a$, $-h \leqslant x_2 \leqslant h$, $0 \leqslant x_3 \leqslant l$. The end $x_3 = l$ is built-in and the beam is bent by a force P applied at the free end $x_3 = 0$ and acting in the x_2 direction. The stress tensor has components

$$(T_{ij}) = \begin{pmatrix} 0 & 0 & 0 \\ 0 & 0 & A + Bx_2^2 \\ 0 & A + Bx_2^2 & Cx_2x_3 \end{pmatrix}$$

where A, B and C are constants. (a) Show that this stress satisfies the equations of equilibrium with no body forces provided $2B + C = 0$; (b) determine the relation between A and B if no traction acts on the sides $x_2 = \pm h$; (c) express the resultant force on the free end $x_3 = 0$ in terms of A, B and C and hence, with (a) and (b), show that $C = -3P/4ah^3$.

5. The stress in the cantilever beam of Problem 4 is now given by

$$(T_{ij}) = \begin{pmatrix} 0 & 0 & 0 \\ 0 & C(\frac{1}{3}x_2^3 - h^2x_2 - \frac{2}{3}h^3) & Cx_3(h^2 - x_2^2) \\ 0 & Cx_3(h^2 - x_2^2) & C(x_2x_3^2 - \frac{2}{3}x_2^3) + Dx_2 \end{pmatrix}$$

where C and D are constants. (a) Show that this stress satisfies the equations of equilibrium with no body forces; (b) show that the traction on the surface $x_2 = -h$ is zero; (c) find the magnitude and direction of the traction on the surface $x_2 = h$, and hence the total force on this surface; (d) find the resultant force on the surface $x_3 = l$. Prove that the traction on this surface exerts zero bending couple on it provided that $C(5l^2 - 2h^2) + 5D = 0$.

6. The stress components in a thin plate bounded by $x_1 = \pm L$ and $x_2 = \pm h$ are given by

$$T_{11} = Wm^2 \cos(\tfrac{1}{2}\pi x_1/L) \sinh mx_2,$$
$$T_{22} = -\tfrac{1}{4}W\pi^2L^{-2} \cos(\tfrac{1}{2}\pi x_1/L) \sinh mx_2$$
$$T_{12} = \tfrac{1}{2}W\pi mL^{-1} \sin(\tfrac{1}{2}\pi x_1/L) \cosh mx_2,$$
$$T_{13} = T_{23} = T_{33} = 0$$

where W and m are constants. (a) Verify that this stress satisfies the equations of equilibrium, with no body forces; (b) find the tractions on the edges $x_2 = h$ and $x_1 = -L$; (c) find the principal stress components and the principal axes of stress at $(0, h, 0)$ and at $(L, 0, 0)$.

7. A solid circular cylinder has radius a and length L, its axis coincides with the x_3-axis, and its ends lie in the planes $x_3=-L$ and $x_3=0$. The cylinder is subjected to axial tension, bending and torsion, such that the stress tensor is given by

$$(T_{ij})=\begin{pmatrix} 0 & 0 & -\alpha x_2 \\ 0 & 0 & \alpha x_1 \\ -\alpha x_2 & \alpha x_1 & \beta+\gamma x_1+\delta x_2 \end{pmatrix}$$

where α, β, γ and δ are constants. (a) Verify that these stress components satisfy the equations of equilibrium with no body forces; (b) verify that no traction acts on the curved surface of the cylinder; (c) find the traction on the end $x_3=0$, and hence show that the resultant force on this end is an axial force of magnitude $\pi a^2\beta$, and that the resultant couple on this end has components $(\frac{1}{4}\pi a^4\delta, -\frac{1}{4}\pi a^4\gamma, \frac{1}{2}\pi a^4\alpha)$ about the x_1-, x_2- and x_3-axes; (d) for the case in which bending is absent $(\gamma=0, \delta=0)$ find the principal stress components. Verify that two of these components are equal on the axis of the cylinder, but that elsewhere they are all different provided that $\alpha\neq 0$. Find the principal stress direction which corresponds to the intermediate principal stress component.

8. A cylinder whose axis is parallel to the x_3-axis and whose normal cross-section is the square $-a\leqslant x_1\leqslant a$, $-a\leqslant x_2\leqslant a$, is subjected to torsion by couples acting over its ends $x_3=0$ and $x_3=L$. The stress components are given by $T_{13}=\partial\psi/\partial x_2$, $T_{23}=-\partial\psi/\partial x_1$, $T_{11}=T_{12}=T_{22}=T_{33}=0$, where $\psi=\psi(x_1, x_2)$. (a) Show that these stress components satisfy the equations of equilibrium with no body forces; (b) show that the difference between the maximum and minimum principal stress components is $2\{(\partial\psi/\partial x_1)^2+(\partial\psi/\partial x_2)^2\}^{\frac{1}{2}}$, and find the principal axis which corresponds to the zero principal stress component; (c) for the special case $\psi=(x_1^2-a^2)(x_2^2-a^2)$ show that the lateral surfaces are free from traction and that the couple acting on each end face is $32a^6/9$.

9. Let $\boldsymbol{n}$ be a unit vector, $\boldsymbol{t}^{(n)}$ the traction on the surface normal to $\boldsymbol{n}$, and S the magnitude of the shear stress on this surface, so that S is the component of $\boldsymbol{t}^{(n)}$ perpendicular to $\boldsymbol{n}$. Prove that as $\boldsymbol{n}$ varies, S has stationary values when $\boldsymbol{n}$ is perpendicular to one of the principal axes of stress, and bisects the angle between the other two. Prove also that the maximum and minimum values of S are $\pm\frac{1}{2}(T_1-T_3)$.

6

Motions and deformations

6.1 Rigid-body motions

We employ the notation introduced in Section 4.1, in which the particles of a body are labelled by their coordinates X_R in a reference configuration at the reference time $t=0$. If at a later time t the particle X_R has coordinates x_i, then the equations

$$x_i = x_i(X_R, t), \qquad \text{or} \qquad \boldsymbol{x} = \boldsymbol{x}(\mathbf{X}, t) \tag{6.1}$$

describe a *motion* of the body; they give the position of each particle at time t. In Chapter 4 we were mainly concerned with the kinematics of individual particles. In this chapter we consider how a particle moves in relation to its neighbouring particles.

In a *rigid-body motion* the body $\mathscr{B}$ moves without changing its shape. The distance between any two particles of $\mathscr{B}$ does not change during a rigid-body motion; neither does the angle between the two lines joining a particle to two other particles.

Translation. A translation is a rigid-body motion of a body in which every particle undergoes the same displacement; thus the motion is described by the equations

$$x_i = X_i + c_i(t), \qquad \text{or} \qquad \boldsymbol{x} = \mathbf{X} + \boldsymbol{c}(t) \tag{6.2}$$

where the vector $\boldsymbol{c}$ is independent of position and depends only on t.

Rotation. Consider a motion in which $\mathscr{B}$ rotates in the anti-clockwise direction through an angle α (which may depend on t) about the x_3-axis. Thus, in Fig. 6.1, the particle initially at a typical point P_0 moves to the point P, such that $NP_0 = NP$ and the angle between NP_0 and NP is α. Then by elementary geometry

$$x_1 = X_1 \cos\alpha - X_2 \sin\alpha, \qquad x_2 = X_1 \sin\alpha + X_2 \cos\alpha, \qquad x_3 = X_3 \tag{6.3}$$

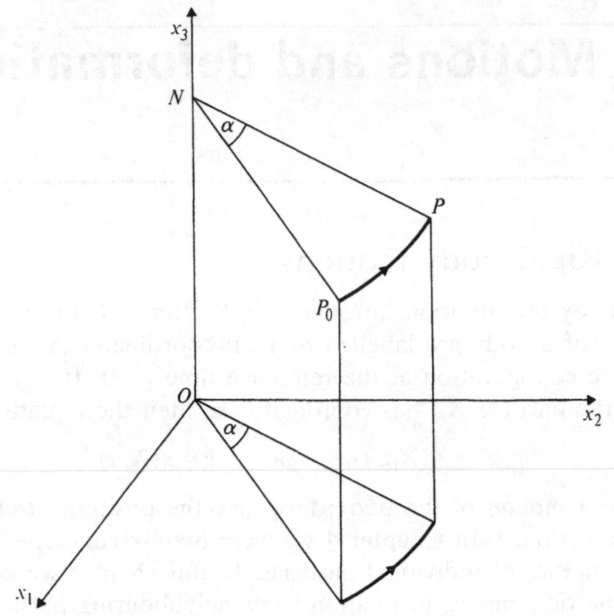

Figure 6.1 Rotation about the x_3-axis

or, in tensor notation

$$\boldsymbol{x} = \boldsymbol{Q} \cdot \mathbf{X} \tag{6.4}$$

where the components, referred to base vectors $\boldsymbol{e}_i$, of the tensor $\boldsymbol{Q}$ are given by

$$(Q_{iR}) = \begin{pmatrix} \cos\alpha & -\sin\alpha & 0 \\ \sin\alpha & \cos\alpha & 0 \\ 0 & 0 & 1 \end{pmatrix} \tag{6.5}$$

It is easily verified that $\boldsymbol{Q}$ is an orthogonal tensor, and so we also have

$$\mathbf{X} = \boldsymbol{Q}^{\mathrm{T}} \cdot \boldsymbol{x} \tag{6.6}$$

Now consider a more general rotation in which $\mathscr{B}$ rotates about an arbitrary axis through the origin O. The direction of the axis is defined by a unit vector $\boldsymbol{n}$, and the angle of rotation is α in the sense of the rotation of a right-handed screw travelling in the direction of $\boldsymbol{n}$. We refer to Fig. 6.2. Let OQ represent the axis of rotation and let $\mathbf{X}$ be the position vector of a typical point P_0 in $\mathscr{B}$. In the rotation, the particle which is initially at P_0 moves

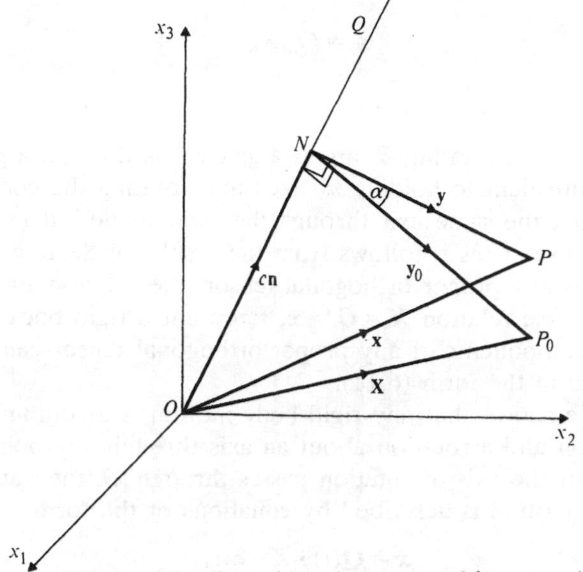

Figure 6.2 Rotation about an arbitrary axis

to P, with position vector $\boldsymbol{x}$. Hence P_0 and P lie in a plane normal to $\boldsymbol{n}$; suppose that this plane intersects OQ at N. Then $NP_0 = NP$, and $\alpha = \angle P_0NP$, and the position vector of N relative to O is $c\boldsymbol{n}$, where, from Fig. 6.2,

$$c = \boldsymbol{n} \cdot \boldsymbol{X} = \boldsymbol{n} \cdot \boldsymbol{x} \tag{6.7}$$

We also denote by $\mathbf{y}_0$ and $\mathbf{y}$ the position vectors of P_0 and P respectively, relative to N. Thus

$$\boldsymbol{X} = c\boldsymbol{n} + \mathbf{y}_0, \qquad \boldsymbol{x} = c\boldsymbol{n} + \mathbf{y} \tag{6.8}$$

Since $\mathbf{y}$ and $\mathbf{y}_0$ have the same magnitude, it follows from Fig. 6.2 that

$$\mathbf{y} = \mathbf{y}_0 \cos\alpha + \boldsymbol{n} \times \mathbf{y}_0 \sin\alpha$$

Hence, from (6.7) and (6.8),

$$\begin{aligned} \boldsymbol{x} &= c\boldsymbol{n} + (\boldsymbol{X} - c\boldsymbol{n})\cos\alpha + \boldsymbol{n} \times (\boldsymbol{X} - c\boldsymbol{n})\sin\alpha \\ &= \boldsymbol{X}\cos\alpha + (\boldsymbol{n} \times \boldsymbol{X})\sin\alpha + c(1 - \cos\alpha)\boldsymbol{n} \\ &= \boldsymbol{X}\cos\alpha + (\boldsymbol{n} \times \boldsymbol{X})\sin\alpha + (\boldsymbol{n} \cdot \boldsymbol{X})(1 - \cos\alpha)\boldsymbol{n} \end{aligned} \tag{6.9}$$

In components, (6.9) may be written as

$$x_i = X_i \cos\alpha + e_{ijR} n_j X_R \sin\alpha + (1 - \cos\alpha) X_R n_R n_i \tag{6.10}$$

or as

$$x_i = Q_{iR}X_R$$

where

$$Q_{iR} = \delta_{iR}\cos\alpha + e_{ijR}n_j\sin\alpha + (1-\cos\alpha)n_i n_R \qquad (6.11)$$

It is evident that rotating $\mathscr{B}$ about a given axis through a given angle is equivalent to holding $\mathscr{B}$ fixed and rotating the coordinate system about the same axis through the same angle but in the opposite sense. Thus it follows from the results of Section 3.2 that, if $\boldsymbol{Q}$ is any proper orthogonal tensor, the relation $\boldsymbol{x} = \boldsymbol{Q}\cdot\boldsymbol{X}$, and the inverse relation $\boldsymbol{X} = \boldsymbol{Q}^{\mathrm{T}}\cdot\boldsymbol{x}$, represent a rigid-body rotation. The components of any proper orthogonal tensor can be represented in the form (6.11).

It can be shown that any rigid-body motion is a combination of a translation and a rotation about an axis through any point. In particular, if the axis of rotation passes through O, then any rigid-body motion is described by equations of the form

$$\boldsymbol{x} = \boldsymbol{Q}(t)\cdot\boldsymbol{X} + \boldsymbol{c}(t)$$

or $\qquad (6.12)$

$$\boldsymbol{X} = \boldsymbol{Q}^{\mathrm{T}}(t)\cdot\boldsymbol{x} + \boldsymbol{c}_1(t)$$

where $\boldsymbol{c}_1(t) = -\boldsymbol{Q}^{\mathrm{T}}(t)\boldsymbol{c}(t)$.

6.2 Extension of a material line element

In a general motion a body will change its shape as well as its position and orientation. A motion in which a change of shape takes place is called a *deformation*; a body which can change its shape is *deformable*, in contrast to a *rigid body* which can only undergo rigid-body motions. One of the main problems in the analysis of deformation is to separate that part of a motion which corresponds to a rigid-body motion from the part which involves deformation.

In a deformation, there are changes in distance between particles, whereas in a rigid-body motion there are no such changes. We therefore begin by examining the extension or stretch of a material line element. Consider a segment P_0Q_0 of a straight line lying in the body $\mathscr{B}$ in its reference configuration, such that P_0Q_0 has length δL and is aligned in the direction of a unit vector $\mathbf{A}$,*

* The use of $\mathbf{A}$ to denote a vector in the reference configuration is another exception to our general rule that vectors are denoted by lower-case letters.

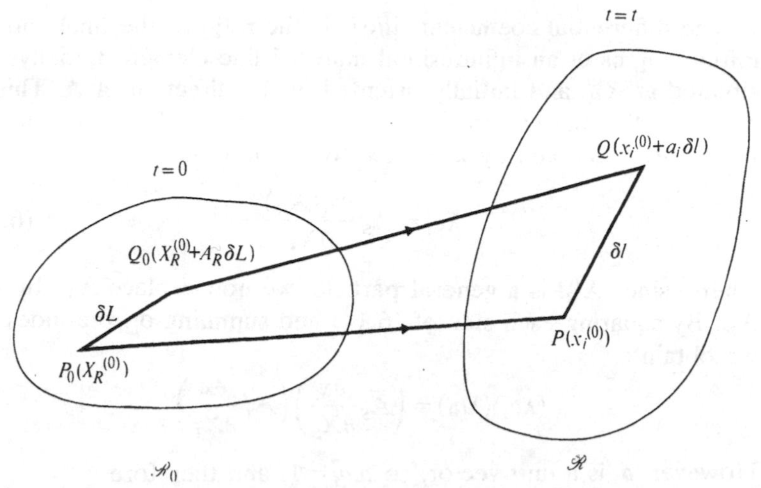

Figure 6.3 Extension of a material line element

as illustrated in Fig. 6.3. Thus if P_0 has coordinates $X_R^{(0)}$, then Q_0 has coordinates $X_R^{(0)} + A_R\,\delta L$. The particles which lie on P_0Q_0 at time $t=0$ form a segment of a *material curve*, and after a motion these particles will in general lie on a new curve in space. The motion is described by the relations (6.1), and we wish to determine the length and orientation of the material line element after the motion. Suppose that t the particles initially at P_0 and Q_0 move to P and Q respectively, and that the line segment PQ has length δl and the direction of a unit vector $\boldsymbol{a}$. Thus if P has coordinates $x_i^{(0)}$, then Q has coordinates $x_i^{(0)} + a_i\,\delta l$. Since P was initially at P_0, it follows from (6.1) that (omitting the argument t)

$$x_i^{(0)} = x_i(X_R^{(0)})$$

and since Q was initially at Q_0, it follows similarly that

$$x_i^{(0)} + a_i\,\delta l = x_i(X_R^{(0)} + A_R\,\delta L)$$

Hence, by Taylor's theorem, since the A_R are of order one,

$$x_i^{(0)} + a_i\,\delta l = x_i(X_R^{(0)}) + A_S\,\delta L\frac{\partial x_i(X_R^{(0)})}{\partial X_S} + O\{(\delta L)^2\}$$

$$= x_i^{(0)} + A_S\,\delta L\frac{\partial x_i(X_R^{(0)})}{\partial X_S} + O\{(\delta L)^2\}$$

Thus, in the limit as $\delta L \to 0$

$$a_i\frac{\mathrm{d}l}{\mathrm{d}L} = A_S\frac{\partial x_i(X_R^{(0)})}{\partial X_S} \qquad (6.13)$$

The differential coefficient dl/dL is the ratio of the final and initial lengths of an infinitesimal material line element initially situated at $X_R^{(0)}$ and initially oriented in the direction of $\mathbf{A}$. This ratio is called the *extension ratio* or *stretch ratio* of the line element and is denoted by λ. Hence (6.13) becomes

$$\lambda a_i = A_S \frac{\partial x_i(X_R)}{\partial X_S} \tag{6.14}$$

where, since $X_R^{(0)}$ is a general particle, we now replace $X_R^{(0)}$ by X_R. By squaring each side of (6.14) and summing on the index i, we obtain

$$(\lambda a_i)(\lambda a_i) = \left(A_S \frac{\partial x_i}{\partial X_S}\right)\left(A_T \frac{\partial x_i}{\partial X_T}\right)$$

However, $\boldsymbol{a}$ is a unit vector, so $a_i a_i = 1$, and therefore

$$\lambda^2 = A_S A_T \frac{\partial x_i}{\partial X_S} \frac{\partial x_i}{\partial X_T} \tag{6.15}$$

When λ is determined from (6.15), the orientation $\boldsymbol{a}$ of the line element in the deformed configuration is then given by (6.14).

If the deformation is described by equations of the form

$$X_R = X_R(x_i, t) \qquad \text{or} \qquad \mathbf{X} = \mathbf{X}(\boldsymbol{x}, t)$$

which give the reference coordinates X_R of the particle which occupies x_i at time t, then in a similar way we may determine the stretch ratio λ, and the orientation $\mathbf{A}$ in the reference configuration, of a line element which has the direction $\boldsymbol{a}$ in the deformed configuration. In essence, it is only necessary to interchange $\mathbf{X}$ and $\boldsymbol{x}$, $\mathbf{A}$ and $\boldsymbol{a}$, and δL and δl, in the above argument. Details are left to the reader (Problem 6.1); the main results are

$$A_S = \lambda a_i \frac{\partial X_S}{\partial x_i} \tag{6.16}$$

$$\lambda^{-2} = a_i a_j \frac{\partial X_S}{\partial x_i} \frac{\partial X_S}{\partial x_j} \tag{6.17}$$

6.3 The deformation gradient tensor

The nine quantities $\partial x_i/\partial X_R$ appeared naturally in the analysis of Section 6.2. They are called the *deformation gradients*. It is clear that these quantities must be involved in the description of how a particle moves in relation to neighbouring particles, and so they are of importance in the analysis of deformation.

We denote

$$F_{iR} = \partial x_i / \partial X_R \tag{6.18}$$

Then F_{iR} are components of a second-order tensor, which is called the *deformation gradient tensor* and is denoted by $\boldsymbol{F}$. To confirm that F_{iR} are components of a tensor, we introduce a new rectangular cartesian coordinate system by a rotation of the axes defined by the orthogonal matrix $\mathbf{M}$. Then in the new system, $\boldsymbol{X}$ and $\boldsymbol{x}$ have components $\bar{X}_R$ and $\bar{x}_i$ respectively, where

$$\bar{X}_R = M_{RS} X_S, \qquad \bar{x}_i = M_{ij} x_j$$
$$X_S = M_{RS} \bar{X}_R, \qquad x_j = M_{ij} \bar{x}_i$$

Then

$$\bar{F}_{iR} = \frac{\partial \bar{x}_i}{\partial \bar{X}_R} = \frac{\partial \bar{x}_i}{\partial x_j} \frac{\partial X_S}{\partial \bar{X}_R} \frac{\partial x_j}{\partial X_S}$$
$$= M_{ij} M_{RS} \frac{\partial x_j}{\partial X_S} = M_{ij} M_{RS} F_{jS}$$

Since the components F_{iR} conform to the tensor transformation law, $\boldsymbol{F}$ is a second-order tensor. In general, $\boldsymbol{F}$ is not a symmetric tensor. By the results of Section 3.4, $\boldsymbol{F}^{\mathrm{T}}$ is also a second-order tensor, and so is $\boldsymbol{F}^{-1}$ provided that $\det \boldsymbol{F} \neq 0$ (we shall show in Section 7.2 that there are physical reasons for assuming that $\det \boldsymbol{F} \neq 0$). Since

$$\frac{\partial x_i}{\partial X_R} \frac{\partial X_R}{\partial x_j} = \frac{\partial x_i}{\partial x_j} = \delta_{ij}$$

$\boldsymbol{F}^{-1}$ is the tensor whose components are F_{Rj}^{-1}, where

$$F_{Rj}^{-1} = \partial X_R / \partial x_j \tag{6.19}$$

The main results of Section 6.2 can now be stated in direct tensor notation. Equation (6.14) may be expressed in the form

$$\boldsymbol{a} = \lambda^{-1} \boldsymbol{F} \cdot \boldsymbol{A} \tag{6.20}$$

and (6.15) as

$$\lambda^2 = \boldsymbol{A} \cdot \boldsymbol{F}^{\mathrm{T}} \cdot \boldsymbol{F} \cdot \boldsymbol{A} \tag{6.21}$$

Similarly, (6.16) and (6.17) may be written, respectively, as

$$\boldsymbol{A} = \lambda \boldsymbol{F}^{-1} \cdot \boldsymbol{a} \tag{6.22}$$

$$\lambda^{-2} = \boldsymbol{a} \cdot (\boldsymbol{F}^{-1})^{\mathrm{T}} \cdot \boldsymbol{F}^{-1} \cdot \boldsymbol{a} \tag{6.23}$$

For the calculation of $\boldsymbol{a}$, $\boldsymbol{A}$ and λ it is often convenient to use matrix notation. If, in a fixed coordinate system, the components of $\boldsymbol{A}$ are written as a column matrix $\mathbf{A}$, those of $\boldsymbol{a}$ as a column matrix $\mathbf{a}$, those of $\boldsymbol{F}$ as a square matrix $\mathbf{F}$, and those of $\boldsymbol{F}^{-1}$ as a square matrix $\mathbf{F}^{-1}$, then (6.20)–(6.23) give

$$\mathbf{a} = \lambda^{-1}\mathbf{F}\mathbf{A}, \qquad \lambda^{2} = \mathbf{A}^{\mathrm{T}}\mathbf{F}^{\mathrm{T}}\mathbf{F}\mathbf{A} \tag{6.24}$$

$$\mathbf{A} = \lambda \mathbf{F}^{-1}\mathbf{a}, \qquad \lambda^{-2} = \mathbf{a}^{\mathrm{T}}(\mathbf{F}^{-1})^{\mathrm{T}}\mathbf{F}^{-1}\mathbf{a} \tag{6.25}$$

If there is no motion, then $x_i = X_i$, $F_{IR} = \delta_{iR}$, and $\boldsymbol{F} = \boldsymbol{I}$.

The components of the displacement vector $\boldsymbol{u}$ are given by $u_i = x_i - X_i$. The *displacement gradients* are

$$\frac{\partial u_i}{\partial X_R} = \frac{\partial x_i}{\partial X_R} - \delta_{iR} = F_{iR} - \delta_{iR} \tag{6.26}$$

and so they are components of the tensor $\boldsymbol{F} - \boldsymbol{I}$. This tensor is called the *displacement gradient tensor.* If there is no motion, then its components are all zero.

Although the tensor $\boldsymbol{F}$ is important in the analysis of deformation, it is not itself a suitable *measure* of deformation. This is because a measure of deformation should have the property that it does not change when no deformation takes place; therefore it must be unchanged in a rigid-body motion. $\boldsymbol{F}$ does not have this property; in fact in the rigid-body motion (6.12), we have $\boldsymbol{F} = \boldsymbol{Q}(t)$.

6.4 Finite deformation and strain tensors

We define a new tensor $\boldsymbol{C}$ as

$$\boldsymbol{C} = \boldsymbol{F}^{\mathrm{T}} \cdot \boldsymbol{F} \tag{6.27}$$

so that the components C_{RS} of $\boldsymbol{C}$ are given by

$$C_{RS} = F_{iR}F_{iS} = \frac{\partial x_i}{\partial X_R}\frac{\partial x_i}{\partial X_S} \tag{6.28}$$

Since $\boldsymbol{C}$ is the inner product of $\boldsymbol{F}^{\mathrm{T}}$ and $\boldsymbol{F}$, it is a second-order tensor; this can also be verified directly by examining the effect of a coordinate transformation on the components C_{RS}. From (6.28) it is evident that $C_{RS} = C_{SR}$, so that $\mathbf{C}$ is a symmetric tensor.

From (6.15) and (6.21) the extension ratio of a material line element with direction $\mathbf{A}$ in the reference configuration is given by

$$\lambda^2 = C_{RS}A_RA_S = \mathbf{A} \cdot \mathbf{C} \cdot \mathbf{A} \tag{6.29}$$

Thus a knowledge of $\mathbf{C}$ enables the extension ratio of any line element to be calculated. Consider an elementary material triangle bounded by three material line elements. Knowledge of the stretch of these line elements completely determines the shape of the triangle (though not its orientation) in a deformed configuration. Hence the components C_{RS} at a particle determine the local deformation in the neighbourhood of that particle.

For the rigid-body motion (6.12), $\mathbf{F} = \mathbf{Q}(t)$ and so

$$\mathbf{C} = \mathbf{Q}^{\mathrm{T}} \cdot \mathbf{Q} = \mathbf{I} \tag{6.30}$$

Hence $\mathbf{C}$ has the constant value $\mathbf{I}$ throughout a rigid-body motion. Thus $\mathbf{C}$ is essentially connected with the deformation, rather than the rigid motion, of a body, and is a suitable measure of the deformation. $\mathbf{C}$ is called the *right Cauchy–Green deformation tensor.*

$\mathbf{C}$ is not a unique measure of deformation. Trivially, any tensor function of $\mathbf{C}$ (such as $\mathbf{C}^2$ or $\mathbf{C}^{-1}$) will serve as such a measure. It is sometimes convenient to employ the measure $\mathbf{C}^{-1}$, which is given in terms of $\mathbf{F}$ by

$$\mathbf{C}^{-1} = \mathbf{F}^{-1} \cdot (\mathbf{F}^{-1})^{\mathrm{T}} \tag{6.31}$$

The components C_{RS}^{-1} of $\mathbf{C}^{-1}$ are given by

$$C_{RS}^{-1} = F_{Ri}^{-1}F_{Si}^{-1} = \frac{\partial X_R}{\partial x_i}\frac{\partial X_S}{\partial x_i} \tag{6.32}$$

Another class of deformation measures is based on the alternative expression (6.17) for λ. If we denote

$$\mathbf{B} = \mathbf{F} \cdot \mathbf{F}^{\mathrm{T}}, \qquad \mathbf{B}^{-1} = (\mathbf{F}^{-1})^{\mathrm{T}} \cdot \mathbf{F}^{-1} \tag{6.33}$$

then $\mathbf{B}$ is the *left Cauchy–Green deformation tensor.* If $\mathbf{B}$ and $\mathbf{B}^{-1}$ have components B_{ij} and B_{ij}^{-1} respectively, then

$$B_{ij} = \frac{\partial x_i}{\partial X_R}\frac{\partial x_j}{\partial X_R}, \qquad B_{ij}^{-1} = \frac{\partial X_R}{\partial x_i}\frac{\partial X_R}{\partial x_j} \tag{6.34}$$

and (6.17) becomes

$$\lambda^{-2} = a_i a_j B_{ij}^{-1} = \boldsymbol{a} \cdot \boldsymbol{B}^{-1} \cdot \boldsymbol{a} \tag{6.35}$$

Hence a knowledge of $\boldsymbol{B}^{-1}$, or equivalently of $\boldsymbol{B}$, is sufficient to determine the local deformation in the neighbourhood of a point in the deformed configuration. It is easy to verify that $\boldsymbol{B} = \boldsymbol{I}$ in a rigid-body motion.

The *Lagrangian strain tensor* $\boldsymbol{\gamma}$ and the *Eulerian strain tensor* $\boldsymbol{\eta}$ are defined by*

$$\boldsymbol{\gamma} = \tfrac{1}{2}(\boldsymbol{C} - \boldsymbol{I}) \tag{6.36}$$

$$\boldsymbol{\eta} = \tfrac{1}{2}(\boldsymbol{I} - \boldsymbol{B}^{-1}) \tag{6.37}$$

Both of these tensors are suitable measures of deformation. They have the properties that $\boldsymbol{\gamma} = \boldsymbol{0}$ and $\boldsymbol{\eta} = \boldsymbol{0}$ in a rigid-body motion; that is, they reduce to zero tensors when there is no deformation.

If the deformation is defined by (6.1), which gives the dependence of $\boldsymbol{x}$ on $\mathbf{X}$, then it is straightforward to calculate $\boldsymbol{F}$ and natural to use $\boldsymbol{C}$, $\boldsymbol{B}$ or $\boldsymbol{\gamma}$ as a deformation measure. The components of these tensors will then be obtained as functions of the material coordinates X_R, and so they describe the deformation in the neighbourhood of a given particle. If the deformation is described by equations which give the dependence of $\mathbf{X}$ on $\boldsymbol{x}$, then it is easier to calculate $\boldsymbol{F}^{-1}$ and the natural deformation measures are $\boldsymbol{C}^{-1}$, $\boldsymbol{B}^{-1}$ and $\boldsymbol{\eta}$; the components of these tensors are obtained as functions of spatial coordinates x_i, and so they describe the deformation which has taken place in the neighbourhood of a given point.

The expressions for the components γ_{RS} of $\boldsymbol{\gamma}$ and η_{ij} of $\boldsymbol{\eta}$ are often given in terms of the displacement gradients. Since

$$\boldsymbol{u} = \boldsymbol{x} - \mathbf{X}$$

we have

$$F_{iR} = \frac{\partial x_i}{\partial X_R} = \frac{\partial u_i}{\partial X_R} + \delta_{iR}$$

Hence, from (6.28) and (6.36),

$$\gamma_{RS} = \frac{1}{2}\left\{\left(\frac{\partial u_i}{\partial X_R} + \delta_{iR}\right)\left(\frac{\partial u_i}{\partial X_S} + \delta_{iS}\right) - \delta_{RS}\right\}$$

$$= \frac{1}{2}\left(\frac{\partial u_R}{\partial X_S} + \frac{\partial u_S}{\partial X_R} + \frac{\partial u_i}{\partial X_R}\frac{\partial u_i}{\partial X_S}\right) \tag{6.38}$$

* The use of $\boldsymbol{\gamma}$ and $\boldsymbol{\eta}$ to denote strain tensors is a departure from our convention of denoting second-order tensors by bold-face italic capital letters.

so that, for example,

$$\gamma_{11}=\frac{\partial u_1}{\partial X_1}+\frac{1}{2}\left\{\left(\frac{\partial u_1}{\partial X_1}\right)^2+\left(\frac{\partial u_2}{\partial X_1}\right)^2+\left(\frac{\partial u_3}{\partial X_1}\right)^2\right\}$$

and

$$\gamma_{23}=\frac{1}{2}\left\{\frac{\partial u_2}{\partial X_3}+\frac{\partial u_3}{\partial X_2}+\frac{\partial u_1}{\partial X_2}\frac{\partial u_1}{\partial X_3}+\frac{\partial u_2}{\partial X_2}\frac{\partial u_2}{\partial X_3}+\frac{\partial u_3}{\partial X_2}\frac{\partial u_3}{\partial X_3}\right\}$$

Similarly,

$$F_{Ri}^{-1}=\frac{\partial X_R}{\partial x_i}=\delta_{Ri}-\frac{\partial u_R}{\partial x_i}$$

and it follows from (6.34) and (6.37) that

$$\eta_{ij}=\frac{1}{2}\left(\frac{\partial u_i}{\partial x_j}+\frac{\partial u_j}{\partial x_i}-\frac{\partial u_R}{\partial x_i}\frac{\partial u_R}{\partial x_j}\right) \tag{6.39}$$

and so, for example,

$$\eta_{11}=\frac{\partial u_1}{\partial x_1}-\frac{1}{2}\left\{\left(\frac{\partial u_1}{\partial x_1}\right)^2+\left(\frac{\partial u_2}{\partial x_1}\right)^2+\left(\frac{\partial u_3}{\partial x_1}\right)^2\right\}$$

$$\eta_{23}=\frac{1}{2}\left\{\frac{\partial u_2}{\partial x_3}+\frac{\partial u_3}{\partial x_2}-\frac{\partial u_1}{\partial x_2}\frac{\partial u_1}{\partial x_3}-\frac{\partial u_2}{\partial x_2}\frac{\partial u_2}{\partial x_3}-\frac{\partial u_3}{\partial x_2}\frac{\partial u_3}{\partial x_3}\right\}$$

The calculation of the deformation and strain tensor components for a given deformation is most easily carried out using matrix operations. We denote

$$\begin{aligned}&\mathbf{F}=(F_{iR})=(\partial x_i/\partial X_R), \qquad \mathbf{F}^{-1}=(F_{Ri}^{-1})=(\partial X_R/\partial x_i),\\ &\mathbf{C}=(C_{RS}), \qquad \mathbf{B}=(B_{ij}), \qquad \mathbf{C}^{-1}=(C_{RS}^{-1}), \qquad \mathbf{B}^{-1}=(B_{ij}^{-1}),\\ &\mathbf{a}=(a_1, a_2, a_3)^{\mathrm{T}}, \qquad \mathbf{A}=(A_1, A_2, A_3)^{\mathrm{T}}\end{aligned} \tag{6.40}$$

Then the principal formulae are

$$\begin{aligned}&\mathbf{C}=\mathbf{F}^{\mathrm{T}}\mathbf{F}, \qquad \mathbf{C}^{-1}=\mathbf{F}^{-1}(\mathbf{F}^{-1})^{\mathrm{T}}, \qquad \mathbf{B}=\mathbf{F}\mathbf{F}^{\mathrm{T}}, \qquad \mathbf{B}^{-1}=(\mathbf{F}^{-1})^{\mathrm{T}}\mathbf{F}^{-1},\\ &(\lambda^2)=\mathbf{A}^{\mathrm{T}}\mathbf{C}\mathbf{A}, \qquad (\lambda^{-2})=\mathbf{a}^{\mathrm{T}}\mathbf{B}^{-1}\mathbf{a},\\ &2(\gamma_{RS})=\mathbf{C}-\mathbf{I}, \qquad 2(\eta_{ij})=\mathbf{I}-\mathbf{B}^{-1}.\end{aligned} \tag{6.41}$$

The tensors $\boldsymbol{C}$, $\boldsymbol{C}^{-1}$, $\boldsymbol{B}$, $\boldsymbol{B}^{-1}$, $\boldsymbol{\gamma}$ and $\boldsymbol{\eta}$ are all symmetric second-order tensors, so they all have real principal components and orthogonal principal directions. Consideration of these is deferred to Chapter 9.

6.5 Some simple finite deformations

(a) *Uniform extensions.* Suppose a body, say a long bar of uniform cross-section, is extended uniformly in the direction of the x_1-axis to a length λ_1 times its original length. Then, if the particle at the origin is fixed in position, $x_1 = \lambda_1 X_1$. This defines a uniform extension in the x_1 direction. If the body undergoes uniform extensions in all three coordinate directions, the deformation is described by the equations

$$x_1 = \lambda_1 X_1, \qquad x_2 = \lambda_2 X_2, \qquad x_3 = \lambda_3 X_3 \tag{6.42}$$

where λ_1, λ_2, λ_3 are constants, or possibly functions of t. Some special cases of (6.42) are of interest. If $\lambda_2 = \lambda_3$, then the body undergoes a uniform expansion or contraction in all directions transverse to the x_1 direction. If $\lambda_1 = \lambda_2 = \lambda_3$, the body undergoes a uniform expansion or contraction in all directions; this is called a *uniform dilation.* If $\lambda_1 = \lambda_2^{-1}$ and $\lambda_3 = 1$, then areas are conserved in planes normal to the x_3 direction, and the deformation is a *pure shear.*

For the deformation (6.42), we readily obtain from (6.40) and (6.41)

$$\mathbf{F} = \begin{pmatrix} \lambda_1 & 0 & 0 \\ 0 & \lambda_2 & 0 \\ 0 & 0 & \lambda_3 \end{pmatrix}, \qquad \mathbf{B} = \mathbf{C} = \begin{pmatrix} \lambda_1^2 & 0 & 0 \\ 0 & \lambda_2^2 & 0 \\ 0 & 0 & \lambda_3^2 \end{pmatrix} \tag{6.43}$$

$$2(\gamma_{RS}) = \begin{pmatrix} \lambda_1^2 - 1 & 0 & 0 \\ 0 & \lambda_2^2 - 1 & 0 \\ 0 & 0 & \lambda_3^2 - 1 \end{pmatrix},$$

$$2(\eta_{ij}) = \begin{pmatrix} 1 - \lambda_1^{-2} & 0 & 0 \\ 0 & 1 - \lambda_2^{-2} & 0 \\ 0 & 0 & 1 - \lambda_3^{-2} \end{pmatrix}$$

(b) *Simple shear.* In this deformation, parallel planes are displaced relative to each other by an amount proportional to the distance between the planes and in a direction parallel to the planes. For example, the simple shear deformation illustrated in Fig. 6.4 is described by the equations

$$x_1 = X_1 + X_2 \tan\gamma, \qquad x_2 = X_2, \qquad x_3 = X_3 \tag{6.44}$$

Here the planes $X_2 =$ constant are the *shear planes* and the X_1

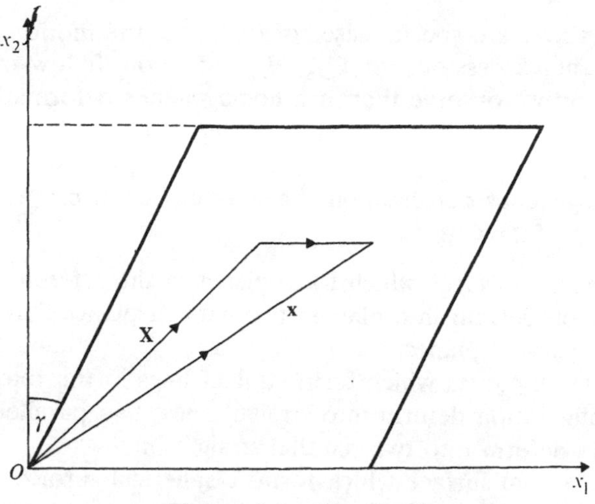

Figure 6.4 Simple shear

direction is the *shear direction.* The angle γ is a measure of the amount of shear. Note that a simple shear involves no change in volume of any portion of the body. For the deformation (6.44), we find from (6.40) and (6.41) that

$$\mathbf{F}=\begin{pmatrix}1 & \tan\gamma & 0\\ 0 & 1 & 0\\ 0 & 0 & 1\end{pmatrix}, \qquad \mathbf{F}^{-1}=\begin{pmatrix}1 & -\tan\gamma & 0\\ 0 & 1 & 0\\ 0 & 0 & 1\end{pmatrix}$$

$$\mathbf{C}=\begin{pmatrix}1 & \tan\gamma & 0\\ \tan\gamma & 1+\tan^2\gamma & 0\\ 0 & 0 & 1\end{pmatrix}, \qquad \mathbf{C}^{-1}=\begin{pmatrix}1+\tan^2\gamma & -\tan\gamma & 0\\ -\tan\gamma & 1 & 0\\ 0 & 0 & 1\end{pmatrix} \quad (6.45)$$

$$\mathbf{B}=\begin{pmatrix}1+\tan^2\gamma & \tan\gamma & 0\\ \tan\gamma & 1 & 0\\ 0 & 0 & 1\end{pmatrix}, \qquad \mathbf{B}^{-1}=\begin{pmatrix}1 & -\tan\gamma & 0\\ -\tan\gamma & 1+\tan^2\gamma & 0\\ 0 & 0 & 1\end{pmatrix}$$

The components of $\boldsymbol{\gamma}$ and $\boldsymbol{\eta}$ follow from (6.41).

(c) *Homogeneous deformations.* These are motions of the form

$$x_i = c_i + A_{iR}X_R$$

or (6.46)

$$\boldsymbol{x} = \boldsymbol{c} + \mathbf{A}\cdot\mathbf{X}$$

where c_i and A_{iR} are constants, or functions of time. Cases (a) and (b) above are special cases of (6.46). In the motion (6.46), $\boldsymbol{F}=\mathbf{A}$. The expressions for C_{RS}, B_{ij}, and so on, follow from (6.41), and we observe that in a homogeneous deformation all the deformation and strain tensors are independent of the coordinates x_i or X_R.

Homogeneous deformations have a number of properties including the following:

(i) Material surfaces which form planes in the reference configuration deform into planes; two parallel planes deform into two parallel planes.

(ii) Material curves which form straight lines in the reference configuration deform into straight lines; two parallel straight lines deform into two parallel straight lines.

(iii) A material surface which forms a spherical surface in the reference configuration is deformed into an ellipsoidal surface.

The proof of these and other similar results is straightforward. As an example we prove (i). The equation satisfied by the material coordinates X_R of particles which initially lie on a plane with unit normal $\boldsymbol{n}$ and perpendicular distance p from the origin is

$$\boldsymbol{n}\cdot\boldsymbol{X}=p$$

After deformation, the same particles lie on a surface such that their position vectors $\boldsymbol{x}$ are related to $\boldsymbol{X}$ by (6.46). Hence

$$\boldsymbol{n}\cdot\mathbf{A}^{-1}\cdot(\boldsymbol{x}-\mathbf{c})=p$$

This is the equation of a plane whose normal is in the direction of the vector $\boldsymbol{n}\cdot\boldsymbol{A}^{-1}$ (it is assumed that $\det\boldsymbol{A}\neq 0$).

(d) *Plane strain.* The deformation defined by

$$x_1=x_1(X_1, X_2), \qquad x_2=x_2(X_1, X_2), \qquad x_3=X_3$$

is called a plane strain. The planes $x_3=$ constant are the deformation planes. Particles which initially lie in a given deformation plane remain in that plane, and their displacement is independent of the X_3 coordinate. Deformations which approximate to plane strain occur in many problems of practical interest.

(e) *Pure torsion.* This deformation is most easily described in terms of cylindrical polar coordinates R, Φ, Z and r, ϕ, z defined

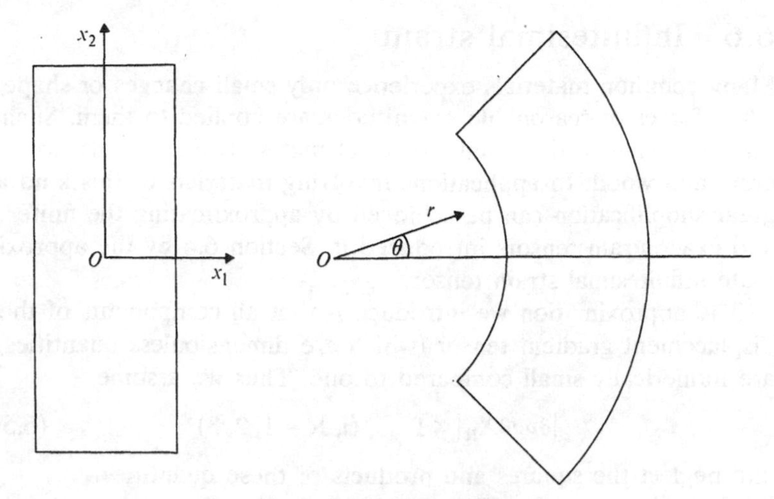

Figure 6.5 Pure flexure

by

$$\begin{aligned} R &= (X_1^2 + X_2^2)^{\frac{1}{2}}, \qquad \Phi = \tan^{-1}(X_2/X_1), \qquad Z = X_3 \\ r &= (x_1^2 + x_2^2)^{\frac{1}{2}}, \qquad \phi = \tan^{-1}(x_2/x_1), \qquad z = x_3 \end{aligned} \tag{6.47}$$

Then a pure torsion is defined by

$$r = R, \qquad \phi = \Phi + \psi Z, \qquad z = Z \tag{6.48}$$

where ψ is constant or a function of time. In this deformation, planes normal to the Z-axis rotate about the Z-axis by an amount which is proportional to Z. The deformation is most easily visualized in terms of the twisting of a circular cylindrical rod whose axis lies along the Z-axis. There are no volume changes and the deformation is not homogeneous.

(f) *Pure flexure.* The deformation illustrated in Fig. 6.5 is described by

$$r = f(X_1), \qquad \phi = g(X_2), \qquad z = X_3 \tag{6.49}$$

This represents the bending of a rectangular block into a sector of a circular cylindrical tube. The material surfaces $X_1 = \text{constant}$, which are parallel planes in the reference configuration, become concentric circular cylindrical surfaces in the deformed configuration, and the material planes $X_2 = \text{constant}$ are deformed from a family of parallel planes into a family of radial planes each containing the z-axis.

6.6 Infinitesimal strain

Many common materials experience only small changes of shape when forces of reasonable magnitudes are applied to them. Such materials include the usual structural materials like metals, concrete and wood. In applications involving materials of this kind a great simplification can be achieved by approximating the finite and exact strain tensors introduced in Section 6.4 by the approximate infinitesimal strain tensor.

The approximation we introduce is that all components of the displacement gradient tensor (which are dimensionless quantities) are numerically small compared to one. Thus we assume

$$|\partial u_i/\partial X_R| \ll 1 \qquad (i, R = 1, 2, 3) \tag{6.50}$$

and neglect the squares and products of these quantities.

Now, since $u_i = x_i - X_i$,

$$\left(\frac{\partial u_i}{\partial x_j}\right) = \left(\delta_{ij} - \frac{\partial X_i}{\partial x_j}\right) = \mathbf{I} - \mathbf{F}^{-1}$$

However, by the binomial expansion,

$$\mathbf{I}-\mathbf{F}^{-1} = \mathbf{I} - \{\mathbf{I} + (\mathbf{F}-\mathbf{I})\}^{-1} = \mathbf{I} - \{\mathbf{I} - (\mathbf{F}-\mathbf{I}) + (\mathbf{F}-\mathbf{I})^2 - (\mathbf{F}-\mathbf{I})^3 + \ldots\}$$

Hence

$$\left(\frac{\partial u_i}{\partial x_j}\right) = (\mathbf{F}-\mathbf{I}) - (\mathbf{F}-\mathbf{I})^2 + (\mathbf{F}-\mathbf{I})^3 - \ldots$$

and so, since $\mathbf{F} - \mathbf{I} = (\partial u_i/\partial X_R)$,

$$\frac{\partial u_i}{\partial x_j} = \frac{\partial u_i}{\partial X_j} - \frac{\partial u_i}{\partial X_R}\frac{\partial u_R}{\partial X_j} + \frac{\partial u_i}{\partial X_R}\frac{\partial u_R}{\partial X_S}\frac{\partial u_S}{\partial X_j} - \ldots \tag{6.51}$$

Therefore, to first order in the displacement gradients, $\partial u_i/\partial x_j \simeq \partial u_i/\partial X_j$, and it is immaterial whether the displacement gradients are formed by differentiation with respect to material coordinates X_R or to spatial coordinates x_i. To this order of approximation, it follows from (6.38) and (6.39) that

$$\gamma_{ij} \simeq \eta_{ij} \simeq \frac{1}{2}\left(\frac{\partial u_i}{\partial X_j} + \frac{\partial u_j}{\partial X_i}\right) \simeq \frac{1}{2}\left(\frac{\partial u_i}{\partial x_j} + \frac{\partial u_j}{\partial x_i}\right) \tag{6.52}$$

The tensor $\boldsymbol{E}$ whose components E_{ij} are defined as

$$E_{ij} = \frac{1}{2}\left(\frac{\partial u_i}{\partial X_j} + \frac{\partial u_j}{\partial X_i}\right) \tag{6.53}$$

is called the *infinitesimal strain tensor.* Thus

$$(E_{ij})=\begin{pmatrix} \dfrac{\partial u_1}{\partial X_1} & \dfrac{1}{2}\left(\dfrac{\partial u_1}{\partial X_2}+\dfrac{\partial u_2}{\partial X_1}\right) & \dfrac{1}{2}\left(\dfrac{\partial u_1}{\partial X_3}+\dfrac{\partial u_3}{\partial X_1}\right) \\ \dfrac{1}{2}\left(\dfrac{\partial u_2}{\partial X_1}+\dfrac{\partial u_1}{\partial X_2}\right) & \dfrac{\partial u_2}{\partial X_2} & \dfrac{1}{2}\left(\dfrac{\partial u_2}{\partial X_3}+\dfrac{\partial u_3}{\partial X_2}\right) \\ \dfrac{1}{2}\left(\dfrac{\partial u_3}{\partial X_1}+\dfrac{\partial u_1}{\partial X_3}\right) & \dfrac{1}{2}\left(\dfrac{\partial u_3}{\partial X_2}+\dfrac{\partial u_2}{\partial X_3}\right) & \dfrac{\partial u_3}{\partial X_3} \end{pmatrix}$$

Both $\boldsymbol{\gamma}$ and $\boldsymbol{\eta}$ reduce to $\boldsymbol{E}$ to the approximation in which squares, products and higher powers of the displacement gradients are neglected. From (6.26) it follows that

$$\boldsymbol{E}=\tfrac{1}{2}(\boldsymbol{F}+\boldsymbol{F}^{\mathrm{T}})-\boldsymbol{I} \tag{6.54}$$

This relation is exact and involves no approximation. Since $\boldsymbol{F}$ is a second-order tensor, $\boldsymbol{E}$ is a second-order tensor, and clearly $\boldsymbol{E}$ is symmetric.

The tensor $\boldsymbol{E}$ cannot be an exact measure of deformation, because it does not remain constant in a rigid-body rotation. To illustrate this, consider the rotation (6.3) through α about the X_3-axis. For this motion we find that

$$(E_{ij})=\begin{pmatrix} -(1-\cos\alpha) & 0 & 0 \\ 0 & -(1-\cos\alpha) & 0 \\ 0 & 0 & 0 \end{pmatrix}$$

Thus E_{11} and E_{22} are not zero. However, they are of second order in the small angle α, and so are neglected in the small displacement gradient approximation.

Although the infinitesimal strain tensor is not an exact measure of deformation, it often provides an excellent approximation to such a measure. Typically, for deformations of structural materials, E_{ij} are of order 0.001 or less, and the approximation neglects this compared with one. The classical theory of linear elasticity, with its numerous successful applications, is constructed on the basis of this approximation. The advantage of the infinitesimal strain tensor is that, unlike γ_{RS} and η_{ij}, the components E_{ij} are linear in the displacement components u_i. This means that the techniques of linear analysis can be applied to the solution of boundary-value problems in, for example, the linear theory of elasticity.

The geometrical interpretation of E_{11} is illustrated in Fig. 6.6. The line element P_0Q_0 of length δL initially lies parallel to the X_1-axis. Since the rotation of the line element is small, its extension, to first order in δL, is

$$u_1(X_1+\delta L, X_2, X_3)-u_1(X_1, X_2, X_3)\simeq\frac{\partial u_1}{\partial X_1}\delta L \qquad (6.55)$$

Hence, to first order, E_{11} is the extension per unit initial length of a line element which is initially parallel to the X_1-axis.

A similar geometrical interpretation of E_{23} is illustrated in Fig. 6.7. Suppose that P_0Q_0 and P_0R_0 are line elements which are initially parallel to the X_2- and X_3-axes. Then, by similar arguments, the angles θ_1 and θ_2 shown in Fig. 6.7 are

$$\theta_1\simeq\frac{\partial u_3}{\partial X_2}, \qquad \theta_2\simeq\frac{\partial u_2}{\partial X_3}. \qquad (6.56)$$

Hence $2E_{23}=\left(\dfrac{\partial u_2}{\partial X_3}+\dfrac{\partial u_3}{\partial X_2}\right)$ is, to first order, the decrease during the deformation in the angle between the initially orthogonal material line elements P_0Q_0 and P_0R_0.

The tensor $\boldsymbol{E}$ possesses the usual properties shared by all symmetric second-order tensors. It has an orthogonal triad of principal axes; if these are chosen as coordinate axes then the matrix of components of $\boldsymbol{E}$ has diagonal form. The corresponding diagonal elements E_1, E_2, E_3 are principal components of infinitesimal

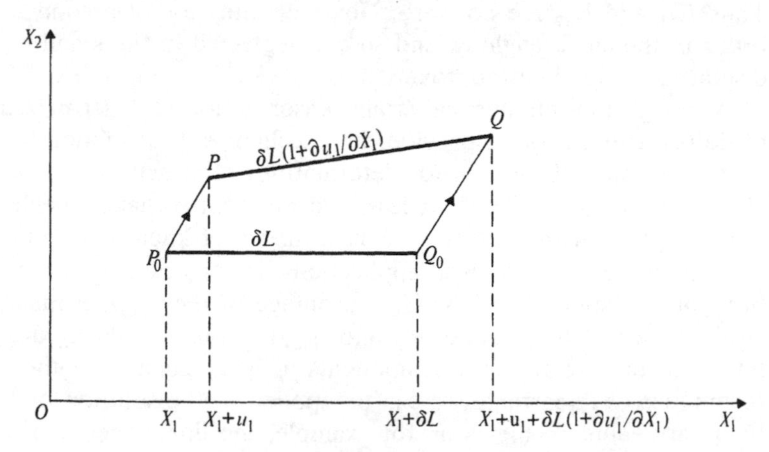

Figure 6.6 Geometrical interpretation of E_{11}

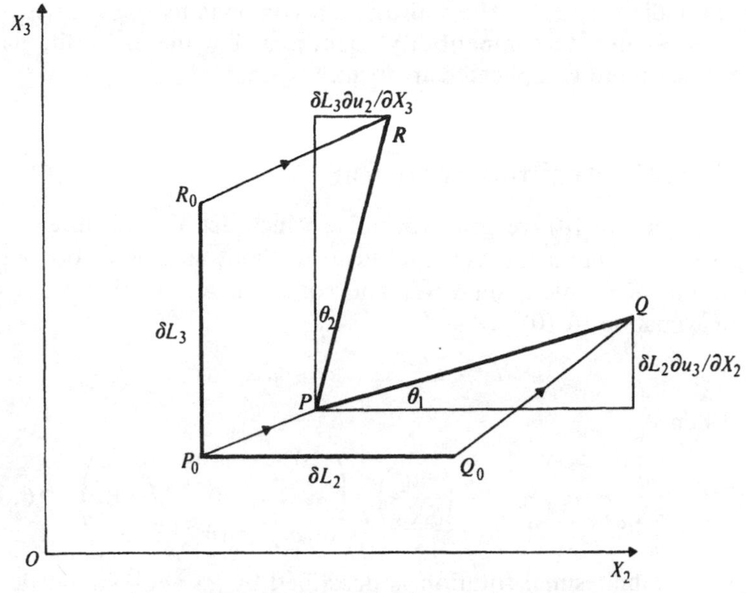

Figure 6.7 Geometrical interpretation of E_{23}

strain. Symmetric functions of E_1, E_2 and E_3 are invariants of the infinitesimal strain tensor.

Because the components E_{ij} are derived from the three displacement components u_i, the E_{ij} are not fully independent, but must satisfy relations obtained by eliminating u_i between them. It can be verified by direct substitution from (6.53) that E_{ij} satisfy the strain compatibility relations

$$K_1 \equiv 2\frac{\partial^2 E_{23}}{\partial X_2\, \partial X_3} - \left(\frac{\partial^2 E_{22}}{\partial X_3^2} + \frac{\partial^2 E_{33}}{\partial X_2^2}\right) = 0 \qquad (6.57)$$

$$L_1 \equiv \frac{\partial^2 E_{11}}{\partial X_2\, \partial X_3} + \frac{\partial}{\partial X_1}\left(\frac{\partial E_{23}}{\partial X_1} - \frac{\partial E_{31}}{\partial X_2} - \frac{\partial E_{12}}{\partial X_3}\right) = 0 \qquad (6.58)$$

and the four similar relations obtained by cyclic permutations of the indices 1, 2, 3. These six compatibility relations are themselves not completely independent, for it can be verified, again by direct substitution, that

$$\frac{\partial K_1}{\partial X_1} = \frac{\partial L_2}{\partial X_3} + \frac{\partial L_3}{\partial X_2} \qquad (6.59)$$

and there are two similar relations obtained by cyclic permutation of the indices 1, 2, 3. The finite strain components γ_{RS} and η_{ij} are also subject to compatibility conditions, but these conditions are much more complicated in form.

6.7 Infinitesimal rotation

In (6.9) and (6.10) we gave formulae which describe a finite rigid-body rotation through the angle α about an axis $\boldsymbol{n}$. For an *infinitesimal* rotation, $\sin\alpha \simeq \alpha$ and $\cos\alpha \simeq 1$, and to this order of approximation (6.10) gives

$$u_i = x_i - X_i = \alpha e_{ijR} n_j X_R$$

and hence

$$\frac{\partial u_i}{\partial X_R} = \alpha e_{ijR} n_j, \qquad \left(\frac{\partial u_i}{\partial X_R}\right) = \begin{pmatrix} 0 & -\alpha n_3 & \alpha n_2 \\ \alpha n_3 & 0 & -\alpha n_1 \\ -\alpha n_2 & \alpha n_1 & 0 \end{pmatrix} \tag{6.60}$$

Thus an infinitesimal rotation is described by an anti-symmetric tensor. We note that this rotation is also described in magnitude and direction by the vector $\alpha\boldsymbol{n}$, and observe the connections between the components of the vector and those of the tensor.

Now consider a general infinitesimal motion with deformation gradient tensor $\boldsymbol{F}$. We define the infinitesimal rotation tensor $\boldsymbol{\Omega}$ and its components Ω_{ij} as follows:

$$\boldsymbol{\Omega} = \tfrac{1}{2}(\boldsymbol{F} - \boldsymbol{F}^{\mathrm{T}}), \qquad \Omega_{ij} = \frac{1}{2}\left(\frac{\partial u_i}{\partial X_j} - \frac{\partial u_j}{\partial X_i}\right)$$

$$(\Omega_{ij}) = \frac{1}{2}\begin{pmatrix} 0 & \dfrac{\partial u_1}{\partial X_2} - \dfrac{\partial u_2}{\partial X_1} & \dfrac{\partial u_1}{\partial X_3} - \dfrac{\partial u_3}{\partial X_1} \\ \dfrac{\partial u_2}{\partial X_1} - \dfrac{\partial u_1}{\partial X_2} & 0 & \dfrac{\partial u_2}{\partial X_3} - \dfrac{\partial u_3}{\partial X_2} \\ \dfrac{\partial u_3}{\partial X_1} - \dfrac{\partial u_1}{\partial X_3} & \dfrac{\partial u_3}{\partial X_2} - \dfrac{\partial u_2}{\partial X_3} & 0 \end{pmatrix} \tag{6.61}$$

Clearly $\boldsymbol{\Omega}$ is a second-order anti-symmetric tensor, and so it can represent an infinitesimal rotation. The displacement gradient tensor $\boldsymbol{F} - \boldsymbol{I}$ is now decomposed into its symmetric and anti-symmetric parts as follows:

$$\boldsymbol{F} - \boldsymbol{I} = \tfrac{1}{2}(\boldsymbol{F} + \boldsymbol{F}^{\mathrm{T}}) - \boldsymbol{I} + \tfrac{1}{2}(\boldsymbol{F} - \boldsymbol{F}^{\mathrm{T}}) = \mathbf{E} + \boldsymbol{\Omega} \tag{6.62}$$

This expresses any infinitesimal motion as the sum of an infinitesimal deformation, represented by $\boldsymbol{E}$, and an infinitesimal rotation, represented by $\boldsymbol{\Omega}$.

The *infinitesimal rotation vector* $\boldsymbol{\omega}$ is defined by

$$\boldsymbol{\omega} = \tfrac{1}{2}\,\mathrm{curl}\,\boldsymbol{u}, \qquad \omega_i = \tfrac{1}{2} e_{ijk}\, \partial u_k/\partial X_j \tag{6.63}$$

Then it follows from (6.61) and (6.63) that

$$\Omega_{jk} = -e_{ijk}\omega_i \tag{6.64}$$

$$\omega_i = -\tfrac{1}{2} e_{ijk}\,\Omega_{jk} \tag{6.65}$$

Further discussion of the rotation will be given in Section 9.2.

The assumption that $\partial u_i/\partial X_R \ll 1$ carries the implication that both the strain and the rotation are small. It is possible to envisage and to realize situations in which the strain components are everywhere small but some material elements undergo large rotations. This may occur, for example, in the bending of a long thin flexible rod. Individual elements of the rod change shape only slightly, but the rotations and displacements can be large. Such problems require careful formulation, and will not be discussed here.

6.8 The rate-of-deformation tensor

In many problems in continuum mechanics the kinematic property of greatest interest is not the change of shape of a body but the rate at which this change is taking place. This is especially the case in fluid mechanics, where it is usually required to find the fluid flow in a particular region of space, and the shape of the body of fluid at a reference time is rarely relevant.

We therefore begin this section by investigating the rate of extension of a material line element; that is, the rate of change of λ for a fixed material line element. The starting point is equation (6.15)

$$\lambda^2 = A_S A_T \frac{\partial x_i}{\partial X_S}\frac{\partial x_i}{\partial X_T} \tag{6.66}$$

which gives λ in terms of material coordinates X_R and the direction cosines A_R of the line element in the reference configuration. It is convenient to begin with (6.66) despite the fact that eventually we wish to express $\mathrm{D}\lambda/\mathrm{D}t$ in terms of spatial coordinates x_i

and the direction cosines a_i of the line element at time t in the current configuration.

We differentiate (6.66) with respect to t, with X_R held constant. Since $Dx_i(X_R, t)/Dt = v_i(X_R, t)$, this gives

$$2\lambda \frac{D\lambda}{Dt} = A_S A_T \left(\frac{\partial x_i}{\partial X_S} \frac{\partial v_i}{\partial X_T} + \frac{\partial x_i}{\partial X_T} \frac{\partial v_i}{\partial X_S} \right) \tag{6.67}$$

To introduce derivatives of v_i with respect to spatial coordinates we use relations of the form

$$\frac{\partial v_i}{\partial X_T} = \frac{\partial v_i}{\partial x_j} \frac{\partial x_j}{\partial X_T}$$

and thereby express (6.67) in the form

$$\lambda \frac{D\lambda}{Dt} = \tfrac{1}{2} A_S A_T \left(\frac{\partial x_i}{\partial X_S} \frac{\partial x_j}{\partial X_T} \frac{\partial v_i}{\partial x_j} + \frac{\partial x_i}{\partial X_T} \frac{\partial x_j}{\partial X_S} \frac{\partial v_i}{\partial x_j} \right)$$

An interchange of the dummy indices i and j in the final term then gives

$$\lambda \frac{D\lambda}{Dt} = \tfrac{1}{2} A_S A_T \frac{\partial x_i}{\partial X_S} \frac{\partial x_j}{\partial X_T} \left(\frac{\partial v_i}{\partial x_j} + \frac{\partial v_j}{\partial x_i} \right)$$

Next we twice employ the relation (6.14) to introduce a_i in place of A_R, and so obtain

$$\lambda^{-1} \frac{D\lambda}{Dt} = \tfrac{1}{2} a_i a_j \left(\frac{\partial v_i}{\partial x_j} + \frac{\partial v_j}{\partial x_i} \right) \tag{6.68}$$

Now $\lambda^{-1} D\lambda/Dt$ is the rate of extension, per unit current length, of a material line element with current direction cosines a_i. For any given direction $\boldsymbol{a}$ this extension rate is, from (6.68), given by $a_i a_j D_{ij}$, where

$$D_{ij} = \frac{1}{2} \left(\frac{\partial v_i}{\partial x_j} + \frac{\partial v_j}{\partial x_i} \right) \tag{6.69}$$

The quantities D_{ij} are the components, referred to base vectors $\boldsymbol{e}_i$, of the *rate-of-deformation tensor* $\boldsymbol{D}$ (other common names are the *rate-of-strain* or *strain-rate tensor*). Note that D_{ij} is linear in the velocity components v_i, and that this linearity is exact and we have not made any approximation in deriving it. We also observe that the right side of (6.68) involves only quantities measured in the current configuration, although we have made use of a reference configuration in order to derive (6.68).

The rate-of-deformation tensor $\boldsymbol{D}$ has properties which in almost every respect (but with an important exception noted below) are analogous to those of the infinitesimal strain tensor $\boldsymbol{E}$. It is readily verified that $\boldsymbol{D}$ is a second-order symmetric tensor. Referred to its principal axes as coordinate axes, the matrix of components of $\boldsymbol{D}$ has diagonal form with principal components D_1, D_2 and D_3. The largest and smallest of the principal components are extremal values of the extension rate for variations of the direction $\boldsymbol{a}$. Symmetric functions of D_1, D_2 and D_3 are invariants of $\boldsymbol{D}$. The components D_{ij} obey compatibility relations which are precisely analogous to the relations (6.57), (6.58) and (6.59) satisfied by E_{ij}, except that differentiation must be with respect to spatial coordinates x_i and these may not be replaced by material coordinates X_R.

The tensor $\boldsymbol{D}$ differs from the tensor $\boldsymbol{E}$ in that it is an *exact* measure of deformation rate, whereas it was emphasized in Section 6.6 that $\boldsymbol{E}$ can never be an exact measure of deformation. The fact that D_{ij} are linear in the velocity components is a fortunate circumstance which simplifies the solution of problems in fluid mechanics.

6.9 The velocity gradient and spin tensors

The deformation-rate tensor $\boldsymbol{D}$ can be identified as the symmetric part of the velocity gradient tensor $\boldsymbol{L}$, whose components L_{ij} are given by

$$L_{ij} = \frac{\partial v_i}{\partial x_j} \tag{6.70}$$

The anti-symmetric part of $\boldsymbol{L}$ is denoted by $\boldsymbol{W}$, and the components of $\boldsymbol{W}$ by W_{ij}, so that

$$W_{ij} = \frac{1}{2}\left(\frac{\partial v_i}{\partial x_j} - \frac{\partial v_j}{\partial x_i}\right) \tag{6.71}$$

and

$$\boldsymbol{L} = \boldsymbol{D} + \boldsymbol{W}, \qquad \boldsymbol{D} = \tfrac{1}{2}(\boldsymbol{L} + \boldsymbol{L}^{\mathrm{T}}), \qquad \boldsymbol{W} = \tfrac{1}{2}(\boldsymbol{L} - \boldsymbol{L}^{\mathrm{T}}) \tag{6.72}$$

It is straightforward to verify that $\boldsymbol{L}$ and $\boldsymbol{W}$ are second-order tensors.

The tensor $\boldsymbol{W}$ is called the *spin* or *vorticity tensor*, and it has properties analogous to those of the infinitesimal rotation tensor, except that no approximation is involved in its derivation or use. It is a measure of the rate of rotation of an element; the expressions (6.72) decompose $\boldsymbol{L}$ into the deformation rate $\boldsymbol{D}$ and the spin $\boldsymbol{W}$. The spin may also be described by the *vorticity vector* $\boldsymbol{w}$, defined by

$$\boldsymbol{w} = \text{curl}\,\boldsymbol{v}, \qquad w_i = e_{ijk}\,\partial v_k/\partial x_j \tag{6.73}$$

By relations similar to (6.64) and (6.65) we have the following connections between $\boldsymbol{W}$ and $\boldsymbol{w}$:

$$W_{jk} = -\tfrac{1}{2}e_{ijk}w_i, \qquad w_i = -e_{ijk}W_{jk} \tag{6.74}$$

In a rigid-body rotation with angular speed ω about an axis through O with unit vector $\boldsymbol{n}$, the velocity is given by

$$\boldsymbol{v} = \omega\boldsymbol{n}\times\boldsymbol{x}, \qquad \text{or} \qquad v_i = e_{ijk}\omega n_j x_k \tag{6.75}$$

Hence, in such a motion, $\boldsymbol{w} = 2\omega\boldsymbol{n}$, and

$$L_{ik} = e_{ijk}\omega n_j, \qquad D_{ik} = 0, \qquad (L_{ik}) = (W_{ik}) = \omega\begin{pmatrix} 0 & -n_3 & n_2 \\ n_3 & 0 & -n_1 \\ -n_2 & n_1 & 0 \end{pmatrix}$$

Thus $\boldsymbol{D}$ vanishes in a rigid-body rotation. Moreover, if a general motion is modified by superposing on it the rigid-body rotation (6.75), then $\boldsymbol{D}$ is the same in the modified motion as it was in the original motion. This confirms that $\boldsymbol{D}$ is unaffected by superposed rotations, and is therefore a suitable measure of the deformation rate.

The material time derivative of F_{iR} is given by

$$\frac{\mathrm{D}}{\mathrm{D}t}(F_{iR}) = \frac{\mathrm{D}}{\mathrm{D}t}\left(\frac{\partial x_i}{\partial X_R}\right) = \frac{\partial v_i}{\partial X_R} = \frac{\partial v_i}{\partial x_j}\frac{\partial x_j}{\partial X_R} = L_{ij}F_{jR}$$

Thus

$$\frac{\mathrm{D}\boldsymbol{F}}{\mathrm{D}t} = \boldsymbol{L}\cdot\boldsymbol{F}, \qquad \text{or} \qquad \boldsymbol{L} = \frac{\mathrm{D}\boldsymbol{F}}{\mathrm{D}t}\cdot\boldsymbol{F}^{-1} \tag{6.76}$$

In the case of small displacement gradients, we have $\boldsymbol{F}^{-1} \simeq \boldsymbol{I}$, and then

$$\boldsymbol{L} \simeq \mathrm{D}\boldsymbol{F}/\mathrm{D}t, \qquad \boldsymbol{D} \simeq \mathrm{D}\boldsymbol{E}/\mathrm{D}t, \qquad \boldsymbol{W} \simeq \mathrm{d}\boldsymbol{\Omega}/\mathrm{D}t, \qquad \boldsymbol{w} \simeq 2\,\mathrm{D}\boldsymbol{\omega}/\mathrm{D}t \tag{6.77}$$

6.10 Some simple flows

(a) *Simple shearing flow.* If the planes $x_2 = \text{constant}$ are the shear planes, and the x_1 direction is the direction of shear, then

$$v_1 = sx_2, \qquad v_2 = 0, \qquad v_3 = 0$$

where s is constant, is a simple shearing flow. The fluid flows in straight lines in the x_1 direction, with speed proportional to its distance from the plane $x_2 = 0$. For this flow

$$D_{12} = \tfrac{1}{2}s, \qquad D_{11} = D_{22} = D_{33} = D_{23} = D_{31} = 0$$
$$W_{12} = \tfrac{1}{2}s, \qquad W_{23} = W_{31} = 0$$

(b) *Rectilinear flow.* In rectilinear flow the material flows in parallel straight lines; this may (but does not always) occur in flow down a pipe of uniform cross-section, or in flow between parallel plates. If the direction of flow is that of the x_3-axis, then

$$v_1 = 0, \qquad v_2 = 0, \qquad v_3 = f(x_1, x_2, x_3)$$

and

$$D_{33} = \partial f/\partial x_3, \qquad D_{13} = D_{31} = \tfrac{1}{2}\,\partial f/\partial x_1, \qquad D_{23} = D_{32} = \tfrac{1}{2}\,\partial f/\partial x_2$$
$$W_{13} = -W_{31} = -\tfrac{1}{2}\,\partial f/\partial x_1, \qquad W_{23} = -W_{32} = -\tfrac{1}{2}\,\partial f/\partial x_2$$

and the remaining components D_{ij} and W_{ij} are zero. If the velocity is independent of x_3, then in addition $D_{33} = 0$.

(c) *Vortex flow.* Flow in the neighbourhood of a vortex line lying along the x_3-axis is described by

$$v_1 = -\frac{\kappa x_2}{x_1^2 + x_2^2}, \qquad v_2 = \frac{\kappa x_1}{x_1^2 + x_2^2}, \qquad v_3 = 0 \qquad (x_1^2 + x_2^2 \neq 0),$$

where κ is a constant. Particles travel in circles around the x_3-axis, with speed inversely proportional to the distance from the axis. The components of $\boldsymbol{D}$ and $\boldsymbol{W}$ are

$$D_{11} = -D_{22} = \frac{2\kappa x_1 x_2}{(x_1^2 + x_2^2)^2}, \qquad D_{12} = \frac{\kappa(x_2^2 - x_1^2)}{(x_1^2 + x_2^2)^2},$$

$$D_{33} = D_{13} = D_{23} = 0 \qquad W_{23} = W_{31} = W_{12} = 0$$

There is a singularity on the vortex line.

(d) *Plane flow.* If the velocity is of the form

$$v_1 = v_1(x_1, x_2, t), \qquad v_2 = v_2(x_1, x_2, t), \qquad v_3 = 0$$

the particles move in planes parallel to $x_3 = 0$, and the velocity is independent of the x_3 coordinate. The non-zero components of $\boldsymbol{D}$ are D_{11}, D_{22} and D_{12}, and these are functions of x_1, x_2 and t only. The only non-zero component of $\boldsymbol{W}$ is $W_{12} = -W_{21}$, and the vorticity vector is in the direction of the x_3-axis. The simple shearing and vortex flows defined above are special cases of plane flow.

6.11 Problems

1. Prove the formulae (6.16) and (6.17).

2. A body undergoes the homogeneous deformation

$$x_1 = \sqrt{2}X_1 + \tfrac{3}{4}\sqrt{2}X_2, \qquad x_2 = -X_1 + \tfrac{3}{4}X_2 + \tfrac{1}{4}\sqrt{2}X_3,$$
$$x_3 = X_1 - \tfrac{3}{4}X_2 + \tfrac{1}{4}\sqrt{2}X_3$$

Find: (a) the direction after the deformation of a line element with direction ratios 1:1:1 in the reference configuration; (b) the stretch of this line element.

3. Find the components of the tensors $\boldsymbol{F}$, $\boldsymbol{C}$, $\boldsymbol{B}$, $\boldsymbol{F}^{-1}$, $\boldsymbol{C}^{-1}$, $\boldsymbol{B}^{-1}$, $\boldsymbol{\gamma}$ and $\boldsymbol{\eta}$ for the deformation

$$x_1 = a_1(X_1 + \alpha X_2), \qquad x_2 = a_2 X_2, \qquad x_3 = a_3 X_3$$

where a_1, a_2, a_3 and α are constants. Find the conditions on these constants for the deformation to be possible in an incompressible material. A body which in the reference configuration is a unit cube with its edges parallel to the coordinate axes undergoes this deformation. Determine the lengths of its edges, and the angles between the edges, after the deformation. Sketch the deformed body.

4. A circular cylinder in its reference configuration has radius A and its axis lies along the X_3-axis. It undergoes the deformation

$$x_1 = \mu\{X_1 \cos(\psi X_3) + X_2 \sin(\psi X_3)\}$$
$$x_2 = \mu\{-X_1 \sin(\psi X_3) + X_2 \cos(\psi X_3)\}, \qquad x_3 = \lambda X_3$$

Find the conditions on the constants λ, μ and ψ for this deformation to be possible in an incompressible material. A line drawn on the surface of the cylinder has unit length and is parallel to the axis of the cylinder in the reference configuration. Find its length after the deformation. Find also the initial length of a line on the surface which has unit length and is parallel to the axis after the deformation.

5. Show that the condition for a material line element to be unchanged in direction during a deformation is $(F_{iR}-\lambda\delta_{iR})A_R=0$. Deduce that the only lines which do not rotate in the simple shear deformation (6.44) are lines which are perpendicular to the X_2-axis. For the deformation

$$x_1=\mu(X_1+X_2\tan\gamma), \qquad x_2=\mu^{-1}X_2, \qquad x_3=X_3 \qquad (\mu\neq 1)$$

show that there are three directions which remain constant. Find these directions and the corresponding stretches.

6. Prove that in the homogeneous deformation (6.46), particles which after the deformation lie on the surface of a sphere of radius b originally lay on the surface of an ellipsoid. Prove that this ellipsoid is a sphere of radius a if $a^2A_{ij}A_{ik}=b^2\delta_{jk}$.

7. A rod of circular cross-section with its axis coincident with the x_3-axis is given a small twist so that its displacement is given by

$$u_1=-\psi x_2x_3, \qquad u_2=\psi x_1x_3, \qquad u_3=0$$

where ψ is constant. Find the components of infinitesimal strain and infinitesimal rotation. Show that one of the principal components of infinitesimal strain is always zero and find the other two principal components. Find also the principal axes of the infinitesimal strain tensor.

8. For the deformation

$$u_1=AX_1+BX_1(X_1^2+X_2^2)^{-1}, \qquad u_2=AX_2+BX_2(X_1^2+X_2^2)^{-1},$$
$$u_3=CX_3$$

where A, B and C are constants, find the components of the tensors $\boldsymbol{F}$, $\boldsymbol{E}$ and $\boldsymbol{\Omega}$. Also find the principal values and principal axes of $\boldsymbol{E}$.

9. For the velocity fields given in Problems 2 and 3 of Chapter 4, find the components of the tensors $\boldsymbol{L}$, $\boldsymbol{D}$ and $\boldsymbol{W}$.

10. Prove that the rate of change of the angle θ between two material line elements whose direction in the current configuration are determined by unit vectors $\boldsymbol{a}$ and $\boldsymbol{b}$ is given by

$$\dot{\theta} \sin \theta = (a_i a_j + b_i b_j) D_{ij} \cos \theta - 2 a_i b_j D_{ij}$$

Deduce that $-2D_{ij}$ $(i \neq j)$ is the rate of change of the angle between two material line elements which instantaneously lie along the x_i- and x_j-axes.

11. An incompressible body is reinforced by embedding in it two families of straight inextensible fibres whose directions in the reference configuration are given by $A_1 = \cos \beta$, $A_2 = \pm \sin \beta$, $A_3 = 0$, where β is constant. The body undergoes the homogeneous deformation

$$x_1 = \mu^{-\frac{1}{2}} \alpha X_1, \qquad x_2 = \mu^{-\frac{1}{2}} \alpha^{-1} X_2, \qquad x_3 = \mu X_3$$

where α and μ are constants. Show that the condition $\lambda = 1$ for inextensibility in the fibre direction requires that $\alpha^2 \cos^2 \beta + \alpha^{-2} \sin^2 \beta = \mu$. Deduce that: (a) the extent to which the body can contract in the x_3 direction is limited by the inequality $\mu \geqslant \sin 2\beta$; (b) when this maximum contraction is achieved, the two families of fibres are orthogonal in the deformed configuration.

Conservation laws

7.1 Conservation laws of physics

Many of the laws of classical physics can be expressed in the form of a statement that some physical quantity is conserved; examples of such quantities are mass, electric charge and momentum. Laws of this kind are general statements and are not restricted in their application to any particular material or class of materials. The mathematical formulations of these laws are therefore equations which must be always satisfied. It is important to distinguish such equations from equations (which we call *constitutive equations*) which describe the properties of particular materials or classes of materials, and which are the subject of Chapters 8 and 10.

We note in passing that the second law of thermodynamics, although it is an important general law of physics, is rather different from the conservation laws mentioned above in that it is expressed as an inequality. Continuum thermodynamics is outside the scope of this introductory text and we shall not discuss it.

7.2 Conservation of mass

The law of conservation of mass will be formulated in two different forms. We first consider the effect of a finite deformation on a volume element.

Deformation of a volume element. The notation of Sections 4.1 and 6.2–6.4 is employed. Consider an elementary tetrahedron in the reference configuration (Fig. 7.1) such that its vertices P_0, Q_0, R_0, S_0 have position vectors $\mathbf{X}^{(0)}$, $\mathbf{X}^{(0)}+\delta\mathbf{X}^{(1)}$, $\mathbf{X}^{(0)}+\delta\mathbf{X}^{(2)}$, $\mathbf{X}^{(0)}+\delta\mathbf{X}^{(3)}$, with coordinates

$$X_R^{(0)}, \qquad X_R^{(0)}+\delta X_R^{(1)}, \qquad X_R^{(0)}+\delta X_R^{(2)}, \qquad X_R^{(0)}+\delta X_R^{(3)} \tag{7.1}$$

respectively. The volume δV of $P_0Q_0R_0S_0$ is

$$\delta V=\tfrac{1}{6}\delta\mathbf{X}^{(1)}\cdot(\delta\mathbf{X}^{(2)}\times\delta\mathbf{X}^{(3)})=\tfrac{1}{6}e_{RST}\,\delta X_R^{(1)}\,\delta X_S^{(2)}\,\delta X_T^{(3)} \tag{7.2}$$

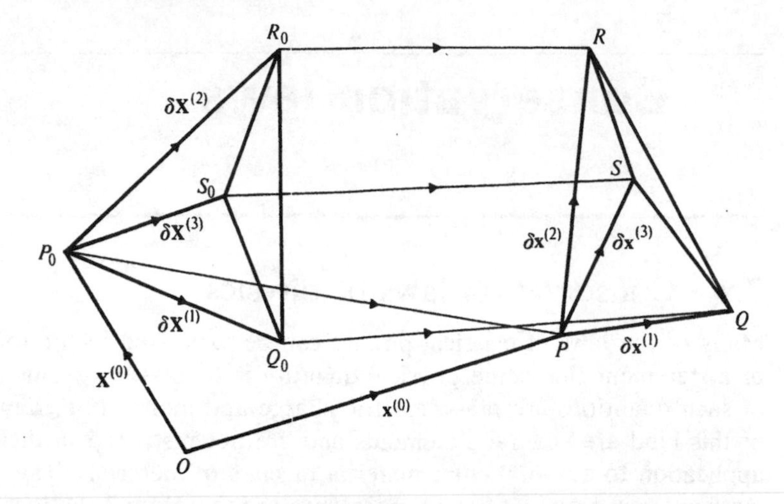

Figure 7.1 Deformation of a volume element

In a deformation the particles initially at P_0, Q_0, R_0, S_0 move to P, Q, R, S with position vectors $\boldsymbol{x}^{(0)}$, $\boldsymbol{x}^{(0)} + \delta\boldsymbol{x}^{(1)}$, etc., and coordinates $x_i^{(0)}$, $x_i^{(0)} + \delta x_i^{(1)}$, etc., respectively. The volume δv of the tetrahedron $PQRS$ is

$$\delta v = \tfrac{1}{6}\delta\boldsymbol{x}^{(1)} \cdot (\delta\boldsymbol{x}^{(2)} \times \delta\boldsymbol{x}^{(3)}) = \tfrac{1}{6}e_{ijk}\, \delta x_i^{(1)}\, \delta x_j^{(2)}\, \delta x_k^{(3)}$$

The deformation is defined by equations of the form $x_i = x_i(X_R, t)$. Hence

$$\delta x_i^{(1)} = \frac{\partial x_i}{\partial X_R}\, \delta X_R^{(1)} + O(\delta X_R^{(1)})^2, \tag{7.3}$$

with the derivatives evaluated at $X_R = X_R^{(0)}$, and similar relations hold for $\delta x_i^{(2)}$ and $\delta x_i^{(3)}$. Therefore the expression for δv becomes

$$\delta v = \tfrac{1}{6}e_{ijk} \frac{\partial x_i}{\partial X_R} \frac{\partial x_j}{\partial X_S} \frac{\partial x_k}{\partial X_T}\, \delta X_R^{(1)}\, \delta X_S^{(2)}\, \delta X_T^{(3)} + O(\delta X_R)^4$$

By using the algebraic result (2.22), this can be written as

$$\delta v = \tfrac{1}{6}e_{RST} \frac{\partial(x_1, x_2, x_3)}{\partial(X_1, X_2, X_3)}\, \delta X_R^{(1)}\, \delta X_S^{(2)}\, \delta X_T^{(3)} + O(\delta X_R^{(p)})^4 \qquad (p = 1, 2, 3) \tag{7.4}$$

where we have introduced the Jacobian

$$\frac{\partial(x_1, x_2, x_3)}{\partial(X_1, X_2, X_3)} = \begin{vmatrix} \partial x_1/\partial X_1 & \partial x_1/\partial X_2 & \partial x_1/\partial X_3 \\ \partial x_2/\partial X_1 & \partial x_2/\partial X_2 & \partial x_2/\partial X_3 \\ \partial x_3/\partial X_1 & \partial x_3/\partial X_2 & \partial x_3/\partial X_3 \end{vmatrix}$$

We now proceed to the limit $\delta X_R^{(p)} \to 0$ $(p = 1, 2, 3)$, so that the initial volume of the tetrahedron tends to zero. Then from (7.2) and (7.4)

$$\frac{dv}{dV} = \frac{\partial(x_1, x_2, x_3)}{\partial(X_1, X_2, X_3)} \tag{7.5}$$

From (6.18) we recognize the above Jacobian as the determinant of the deformation gradient tensor $\boldsymbol{F}$, so that (7.5) can be written as

$$\frac{dv}{dV} = \det \boldsymbol{F} \tag{7.6}$$

If the material is incompressible, then $dv/dV = 1$, and hence $\det \boldsymbol{F} = 1$.

By expanding $\det \boldsymbol{F}$, we obtain

$$\det \boldsymbol{F} = \det \left(\delta_{iR} + \frac{\partial u_i}{\partial X_R} \right) = 1 + \frac{\partial u_i}{\partial X_i} + O\left(\frac{\partial u_i}{\partial X_R} \right)^2$$

Hence, in the case of small displacement gradients,

$$\frac{dv}{dV} = \det \boldsymbol{F} \approx 1 + \frac{\partial u_i}{\partial X_i} = 1 + E_{ii} \tag{7.7}$$

The quantity E_{ii} is called the *dilatation* and is denoted by Δ. From (7.7), Δ is the trace of the infinitesimal strain tensor and so is the first invariant of that tensor. Thus

$$\Delta = E_{ii} = \operatorname{tr} \boldsymbol{E} = E_1 + E_2 + E_3$$

For small deformations, Δ is a measure of the change of volume per unit initial volume of an element.

Conservation of mass – Lagrangian form. Now suppose that the material in the volume element $P_0Q_0R_0S_0$ has mass δm in the reference configuration. Conservation of mass requires that the mass of the material in the material volume element remains constant during the deformation. Hence the initial and final densities, which we denote by ρ_0 and ρ respectively, are

$$\rho_0 = \lim_{\delta V \to 0} \frac{\delta m}{\delta V}, \qquad \rho = \lim_{\delta v \to 0} \frac{\delta m}{\delta v}$$

Hence

$$\frac{\rho_0}{\rho} = \frac{dv}{dV} = \det \boldsymbol{F} \tag{7.8}$$

and this is the required statement of the law of conservation of mass. We note that (7.8) justifies the assumption which was made in Section 6.3 that $\det \boldsymbol{F} \neq 0$, for if $\det \boldsymbol{F} = 0$ then the density is either zero in the initial configuration or infinite in the deformed configuration.

Conservation of mass – Eulerian form. Equation (7.8) expresses the law of conservation of mass in terms of deformation gradients. For many purposes it is more convenient to express the law in terms of the velocity components. For this we consider an arbitrary region $\mathcal{R}$ with surface $\mathcal{S}$, fixed in space in relation to a fixed frame of reference (see Fig. 7.2). The mass conservation law is expressed in the form that the rate at which the mass contained in $\mathcal{R}$ increases is equal to the rate at which mass flows into $\mathcal{R}$ over $\mathcal{S}$. The rate at which mass flows over an element of surface, of area dS, is ρ dS multiplied by the normal component of velocity. Hence

$$\iiint_{\mathcal{R}} \frac{\partial \rho}{\partial t} \mathrm{d}V = -\iint_{\mathcal{S}} \rho \boldsymbol{v} \cdot \boldsymbol{n} \, \mathrm{d}S \tag{7.9}$$

where $\partial\rho/\partial t$ is the rate of increase of ρ at a fixed point in $\mathcal{R}$. The negative sign on the right-hand side appears because $\boldsymbol{n}$ denotes

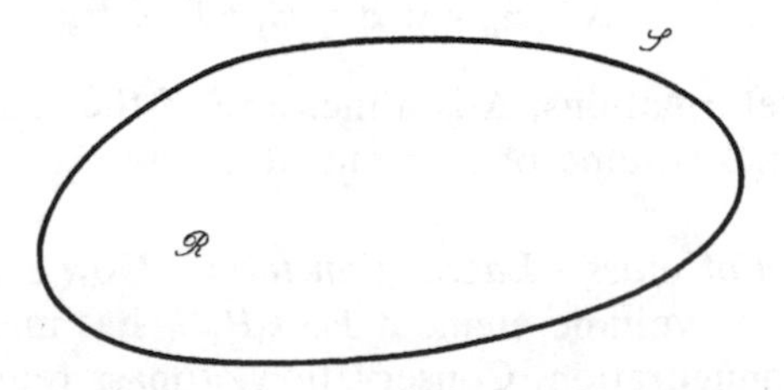

Figure 7.2 The region $\mathcal{R}$

the *outward* normal to $\mathcal{S}$. By applying the divergence theorem to the surface integral, we obtain from (7.9)

$$\iiint_{\mathcal{R}} \left\{\frac{\partial \rho}{\partial t} + \operatorname{div}(\rho \boldsymbol{v})\right\} dV = 0 \tag{7.10}$$

Since the region $\mathcal{R}$ is arbitrary, the integrand in (7.10) must be zero everywhere, for otherwise it would be possible to construct a region for which (7.10) was violated. Hence

$$\frac{\partial \rho}{\partial t} + \operatorname{div}(\rho \boldsymbol{v}) = 0 \tag{7.11}$$

This equation is often called the *continuity equation.* By introducing the components of $\boldsymbol{v}$ and $\boldsymbol{x}$, (7.11) is readily expressed in the following equivalent forms:

$$\frac{\partial \rho}{\partial t} + \boldsymbol{v} \cdot \operatorname{grad} \rho + \rho \operatorname{div} \boldsymbol{v} = 0 \tag{7.12}$$

$$\frac{\partial \rho}{\partial t} + v_i \frac{\partial \rho}{\partial x_i} + \rho \frac{\partial v_i}{\partial x_i} = 0 \tag{7.13}$$

$$\frac{D\rho}{Dt} + \rho \frac{\partial v_i}{\partial x_i} = 0 \tag{7.14}$$

where, as in Section 4.3, $D\rho/Dt$ denotes the material derivative of ρ,

$$\frac{D\rho}{Dt} = \frac{\partial \rho}{\partial t} + v_i \frac{\partial \rho}{\partial x_i}$$

If the material is *incompressible,* then ρ is constant at any particle, so that $D\rho/Dt = 0$. It therefore follows from (7.14) that the incompressibility condition can be expressed in any of the following equivalent forms:

$$\partial v_i/\partial x_i = 0, \qquad \operatorname{div} \boldsymbol{v} = 0 \quad D_{ii} = 0, \qquad \operatorname{tr} \boldsymbol{D} = 0 \tag{7.15}$$

The device of converting a surface integral into a volume integral by the use of the divergence theorem will be used frequently in this chapter. Naturally the results of doing this are valid only if the conditions for the theorem to be applicable are satisfied. The most important of these is that the integrand of the surface integral should be differentiable, and therefore continuous.

Problems do arise in continuum mechanics in which density, velocity, stress and other variables are discontinuous across certain surfaces, which may be stationary or in motion. This situation arises particularly in stress-wave propagation problems. It is not difficult to extend the theory to deal with such cases, and for some problems it is essential to do so. However, in this text it is always assumed that necessary smoothness conditions are satisfied.

7.3 The material time derivative of a volume integral

Suppose that Φ is some physical quantity (such as mass or energy) associated with the particles of a body, and ϕ is the amount of Φ per unit mass. Then the amount of Φ per unit volume is $\rho\phi$ and the amount of Φ contained in a fixed region $\mathscr{R}$ at a given time t is

$$\iiint_{\mathscr{R}} \phi\rho \, \mathrm{d}V \tag{7.16}$$

evaluated at t. In an increment of time δt, the value of ϕ at a given point or at a given particle in $\mathscr{R}$ will (in general) change, and some particles will travel across the surface $\mathscr{S}$ of $\mathscr{R}$, transporting Φ with them. The rate of change of the amount of Φ which is associated with the particles which instantaneously occupy $\mathscr{R}$ at t is called the *material time derivative* of the integral (7.16) and is denoted as

$$\frac{\mathrm{D}}{\mathrm{D}t} \iiint_{\mathscr{R}} \phi\rho \, \mathrm{d}V \tag{7.17}$$

The rate of increase of the amount of Φ within the fixed region $\mathscr{R}$ is equal to the sum of the rate of increase of Φ associated with the particles instantaneously within $\mathscr{R}$, together with the net rate of influx of Φ into $\mathscr{R}$. Thus

$$\iiint_{\mathscr{R}} \frac{\partial(\phi\rho)}{\partial t} \mathrm{d}V = \frac{\mathrm{D}}{\mathrm{D}t} \iiint_{\mathscr{R}} \phi\rho \, \mathrm{d}V - \iint_{\mathscr{S}} \phi\rho \boldsymbol{n} \cdot \boldsymbol{v} \, \mathrm{d}S$$

By applying the divergence theorem to the surface integral, and rearranging, we obtain

$$\frac{\mathrm{D}}{\mathrm{D}t}\iiint_{\mathscr{R}} \phi\rho \,\mathrm{d}V = \iiint_{\mathscr{R}} \left\{\frac{\partial(\phi\rho)}{\partial t} + \operatorname{div}(\phi\rho\boldsymbol{v})\right\} \mathrm{d}V \tag{7.18}$$

If $\phi = 1$, the integral (7.16) represents the mass within $\mathscr{R}$, and conservation of mass requires that the material time derivative of this integral is zero. Hence the integral on the right side of (7.18) (with $\phi = 1$) must have the value zero for all regions $\mathscr{R}$, and so the integrand on the right side is zero. Thus we again obtain the continuity equation in the form (7.11).

For a general quantity ϕ, the integrand of the right side of (7.18) may be written as

$$\phi\left\{\frac{\partial\rho}{\partial t} + \operatorname{div}(\rho\boldsymbol{v})\right\} + \rho\left(\frac{\partial\phi}{\partial t} + \boldsymbol{v}\cdot\operatorname{grad}\phi\right) \tag{7.19}$$

However, by (4.20) and the continuity equation (7.11), the expression (7.19) is just $\rho\,\mathrm{D}\phi/\mathrm{D}t$. Hence (7.18) takes the form

$$\frac{\mathrm{D}}{\mathrm{D}t}\iiint_{\mathscr{R}} \phi\rho \,\mathrm{d}V = \iiint_{\mathscr{R}} \frac{\mathrm{D}\phi}{\mathrm{D}t}\rho \,\mathrm{d}V \tag{7.20}$$

7.4 Conservation of linear momentum

The law of conservation of linear momentum for a particle of mass m states that the rate of change of its linear momentum is equal to the resultant force $\boldsymbol{p}$ applied to it. Thus

$$\frac{\mathrm{D}}{\mathrm{D}t}(m\boldsymbol{v}) = \boldsymbol{p}$$

For a continuum, this statement is generalized as follows: the rate of change of linear momentum of the particles which instantaneously lie within a fixed region $\mathscr{R}$ is proportional to the resultant force applied to the material occupying $\mathscr{R}$. This resultant force consists of the resultant of the body forces $\boldsymbol{b}$ per unit mass acting on the particles in $\mathscr{R}$, together with the resultant of the surface tractions $\boldsymbol{t}^{(n)}$ acting on the surface of $\mathcal{R}$. Hence the law is expressed in the form

$$\frac{\mathrm{D}}{\mathrm{D}t}\iiint_{\mathcal{R}} \rho \boldsymbol{v} \,\mathrm{d}V = \iiint_{\mathcal{R}} \rho \boldsymbol{b} \,\mathrm{d}V + \iint_{\mathcal{S}} \boldsymbol{t}^{(n)} \,\mathrm{d}S \qquad (7.21)$$

In components, after making use of (5.9), this takes the form

$$\frac{\mathrm{D}}{\mathrm{D}t}\iiint_{\mathcal{R}} \rho v_j \,\mathrm{d}V = \iiint_{\mathcal{R}} \rho b_j \,\mathrm{d}V + \iint_{\mathcal{S}} T_{ij} n_i \,\mathrm{d}S$$

where $\boldsymbol{n}$ is the outward normal to $\mathcal{S}$.

We now use (7.20) with ϕ replaced by v_j, and apply the divergence theorem to the surface integral. This gives

$$\iiint_{\mathcal{R}} \left\{\rho \frac{\mathrm{D}v_j}{\mathrm{D}t} - \rho b_j - \frac{\partial T_{ij}}{\partial x_i}\right\} \mathrm{d}V = 0$$

By the usual argument, the integrand is zero, and $\mathrm{D}v_j/\mathrm{D}t = f_j$, where $\boldsymbol{f}$ is the acceleration vector. Hence

$$\frac{\partial T_{ij}}{\partial x_i} + \rho b_j = \rho f_j \qquad (7.22)$$

This is the equation of motion for a continuum. It reduces to the equilibrium equation (5.23) when there is no acceleration.

7.5 Conservation of angular momentum

For a particle, the law of conservation of angular momentum states that

$$\frac{\mathrm{D}}{\mathrm{D}t}\{m(\boldsymbol{x} \times \boldsymbol{v})\} = \boldsymbol{x} \times \boldsymbol{p}$$

where $\boldsymbol{p}$ is the resultant applied force and $\boldsymbol{x}$ is the position vector from an arbitrarily chosen origin. The generalization for a continuum, analogous to (7.21), is

$$\frac{\mathrm{D}}{\mathrm{D}t}\iiint_{\mathcal{R}} \rho \boldsymbol{x} \times \boldsymbol{v} \,\mathrm{d}V = \iiint_{\mathcal{R}} \rho \boldsymbol{x} \times \boldsymbol{b} \,\mathrm{d}V + \iint_{\mathcal{S}} \boldsymbol{x} \times \boldsymbol{t}^{(n)} \,\mathrm{d}S$$

or, in components,

$$\frac{\mathrm{D}}{\mathrm{D}t}\iiint_{\mathcal{R}} \rho e_{ijk}x_j v_k \,\mathrm{d}V = \iiint_{\mathcal{R}} \rho e_{ijk}x_j b_k \,\mathrm{d}V + \iint_{\mathcal{S}} e_{ijk}x_j T_{pk}n_p \,\mathrm{d}S \tag{7.23}$$

In the usual manner, we employ (7.20) with $\phi = e_{ijk}x_j v_k$, transform the surface integral to a volume integral, and equate the integrands of the resulting volume integrals on the two sides of the equation. This gives

$$e_{ijk}\rho\frac{\mathrm{D}}{\mathrm{D}t}(x_j v_k) = e_{ijk}\left\{\rho x_j b_k + \frac{\partial}{\partial x_p}(x_j T_{pk})\right\} \tag{7.24}$$

Now

$$\frac{\mathrm{D}}{\mathrm{D}t}(x_j v_k) = x_j f_k + v_j v_k$$

and

$$\frac{\partial}{\partial x_p}(x_j T_{pk}) = \delta_{pj}T_{pk} + x_j\frac{\partial T_{pk}}{\partial x_p} = T_{jk} + x_j\frac{\partial T_{pk}}{\partial x_p}$$

Hence equation (7.24) can be written as

$$e_{ijk}\left\{T_{jk} + x_j\left(\frac{\partial T_{pk}}{\partial x_p} + \rho b_k - \rho f_k\right) - \rho v_j v_k\right\} = 0 \tag{7.25}$$

However, $e_{ijk}v_j v_k = 0$, and the expression multiplied by x_j in (7.25) is zero by the equation of motion, and so (7.25) reduces to

$$e_{ijk}T_{jk} = 0, \qquad \text{or} \qquad T_{ij} = T_{ji} \tag{7.26}$$

Thus the law of conservation of angular momentum leads to the conclusion that the stress tensor is a symmetric tensor.

It should be mentioned that in writing down (7.23) it is implicitly assumed that no distributed body or surface couples act on the material in $\mathcal{R}$. If such body or surface couples do act, then in general the symmetry of $\boldsymbol{T}$ no longer obtains. However, body and surface couples are of importance only in rather specialized applications, and we shall not consider them.

7.6 Conservation of energy

The *kinetic energy* K of the material which instantaneously occupies a fixed region $\mathscr{R}$ is defined to be

$$K = \tfrac{1}{2}\iiint_{\mathscr{R}} \rho v_i v_i \, \mathrm{d}V \tag{7.27}$$

This is the natural extension to a continuum of the usual expression for the kinetic energy of a particle or rigid body.

The kinetic energy of a continuum is only part of its energy. The remainder is called the *internal energy* E, which is expressed in terms of the *internal energy density* e by

$$E = \iiint_{\mathscr{R}} \rho e \, \mathrm{d}V \tag{7.28}$$

The statement we adopt of the law of conservation of energy is as follows: the material time derivative of $K+E$ is equal to the sum of the rate at which mechanical work is done by the body and surface forces acting on $\mathscr{R}$ and the rate at which other energy enters $\mathscr{R}$.

The 'other energy' may take many different forms. The most important is energy due to heat flux across $\mathscr{S}$. Other possible forms are energy arising from chemical changes inside $\mathscr{R}$, energy arriving by radiation, electromagnetic energy, and so on. We shall consider only the heat flux.

The above statement of the law is not particularly helpful on its own because it can be regarded as being merely a definition of E. It really only becomes useful when some further properties of E or e are specified. To do this leads into the consideration of constitutive equations, which we defer until Chapters 8 and 10.

If q_i denote the components of the heat-flux vector $\boldsymbol{q}$ (that is, $\boldsymbol{q} \cdot \boldsymbol{n}$ is the amount of heat flowing in the sense of the unit vector $\boldsymbol{n}$ across a surface normal to $\boldsymbol{n}$, per unit area per unit time), then the mathematical formulation of the law in the form stated above is

$$\frac{\mathrm{D}}{\mathrm{D}t}\iiint_{\mathscr{R}} \rho(\tfrac{1}{2}v_i v_i + e)\, \mathrm{d}V = \iiint_{\mathscr{R}} \rho b_i v_i \, \mathrm{d}V + \iint_{\mathscr{S}} (T_{ji} v_i - q_j) n_j \, \mathrm{d}S \tag{7.29}$$

The negative sign in the last term arises because $\boldsymbol{n}$ is the *outward* normal to $\mathscr{S}$, and we require the *influx* of heat on the right of the equation. By employing (7.20) on the left side, transforming the surface integral to a volume integral, and equating the integrands, it follows from (7.29), by the argument which is now standard, that

$$\rho\frac{\mathrm{D}}{\mathrm{D}t}(\tfrac{1}{2}v_iv_i+e)=\rho b_iv_i+\frac{\partial}{\partial x_j}(T_{ji}v_i-q_j) \tag{7.30}$$

Now $\mathrm{D}v_i/\mathrm{D}t=f_i$. Hence, after rearrangement, (7.30) becomes

$$-v_i\left(\frac{\partial T_{ji}}{\partial x_j}+\rho b_i-\rho f_i\right)+\rho\frac{\mathrm{D}e}{\mathrm{D}t}-T_{ji}\frac{\partial v_i}{\partial x_j}+\frac{\partial q_j}{\partial x_j}=0$$

The expression in brackets is zero, by the equation of motion (7.22), and so

$$\rho\frac{\mathrm{D}e}{\mathrm{D}t}=T_{ji}\frac{\partial v_i}{\partial x_j}-\frac{\partial q_j}{\partial x_j} \tag{7.31}$$

By interchanging the dummy indices i and j, we have $T_{ji}\,\partial v_i/\partial x_j=T_{ij}\,\partial v_j/\partial x_i$, and, since $\boldsymbol{T}$ is symmetric, $T_{ji}\,\partial v_i/\partial x_j=T_{ij}\,\partial v_i/\partial x_j$. Hence by (6.69),

$$T_{ji}\frac{\partial v_i}{\partial x_j}=\tfrac{1}{2}T_{ij}\left(\frac{\partial v_i}{\partial x_j}+\frac{\partial v_j}{\partial x_i}\right)=T_{ij}D_{ij}$$

and (7.31) may be written as

$$\rho\frac{\mathrm{D}e}{\mathrm{D}t}=T_{ij}D_{ij}-\frac{\partial q_i}{\partial x_i} \tag{7.32}$$

This is the energy equation for a continuum. The term $T_{ij}D_{ij}$ can be interpreted as the rate of working of the stress.

To make further progress it is necessary to assign further properties to e and q. For example, it is often assumed that a gas has a *caloric equation of state*, $e=e(\rho,T)$, where T is temperature. The heat flux $\boldsymbol{q}$ is often assumed to obey *Fourier's law of heat conduction*,

$$\boldsymbol{q}=-\kappa\ \mathrm{grad}\ T \tag{7.33}$$

where κ is the thermal conductivity. Such statements are not general laws, but are particular to certain materials, and are certainly not universally true.

7.7 The principle of virtual work

The principle of virtual work has many applications in continuum mechanics. Although it is not a conservation law, it is convenient to introduce it here. Suppose there is defined in the region $\mathcal{R}$ a stress field with components T_{ij} which satisfy the equilibrium equations

$$\frac{\partial T_{ij}}{\partial x_i} + \rho b_j = 0$$

Also suppose to be defined in $\mathcal{R}$ a velocity field with components v_i which are differentiable with respect to x_i, and let

$$D_{ij} = \frac{1}{2}\left(\frac{\partial v_i}{\partial x_j} + \frac{\partial v_j}{\partial x_i}\right)$$

be the components of the deformation-rate tensor derived from the velocity field v_i.

It is emphasized that T_{ij} and v_i need be in no way connected; T_{ij} may be any equilibrium stress field and v_i any differentiable velocity field.

We form the product $T_{ij}D_{ij}$ and integrate it over the region $\mathcal{R}$. Then, using (5.23) and the symmetry relations $T_{ij} = T_{ji}$, we have

$$\begin{aligned}\iiint_{\mathcal{R}} T_{ij}D_{ij}\,\mathrm{d}V &= \frac{1}{2}\iiint_{\mathcal{R}} T_{ij}\left(\frac{\partial v_i}{\partial x_j} + \frac{\partial v_j}{\partial x_i}\right)\mathrm{d}V \\ &= \iiint_{\mathcal{R}} T_{ij}\frac{\partial v_j}{\partial x_i}\,\mathrm{d}V \\ &= \iiint_{\mathcal{R}} \left\{\frac{\partial}{\partial x_i}(T_{ij}v_j) - v_j\frac{\partial T_{ij}}{\partial x_i}\right\}\mathrm{d}V \\ &= \iiint_{\mathcal{R}} \left\{\frac{\partial}{\partial x_i}(T_{ij}v_j) + \rho v_j b_j\right\}\mathrm{d}V\end{aligned}$$

Finally, by an application of the divergence theorem, we obtain

$$\begin{aligned}\iiint_{\mathcal{R}} T_{ij}D_{ij}\,\mathrm{d}V &= \iint_{\mathcal{S}} T_{ij}v_j n_i\,\mathrm{d}S + \iiint_{\mathcal{R}} \rho v_j b_j\,\mathrm{d}V \\ &= \iint_{\mathcal{S}} \boldsymbol{t}^{(n)}\cdot\boldsymbol{v}\,\mathrm{d}S + \iiint_{\mathcal{R}} \rho\boldsymbol{v}\cdot\boldsymbol{b}\,\mathrm{d}V \qquad (7.34)\end{aligned}$$

where n_i are the direction cosines of the outward normal to the surface $\mathscr{S}$ of $\mathscr{R}$ and $\boldsymbol{t}^{(n)}$ is the surface-traction vector on $\mathscr{S}$ which corresponds to the stress components T_{ij}.

Equation (7.34) is the mathematical expression of the principle of virtual work for a continuum. It states that the rate of working of the stress field T_{ij} in the velocity field v_i is equal to the sum of the rates of working of the surface and body forces associated with T_{ij} in the same field.

An identical argument may be followed with v_i replaced by infinitesimal displacement components u_i and D_{ij} replaced by the infinitesimal strain components E_{ij}.

The relation (7.34) and its analogue in terms of infinitesimal displacement and strain form the basis of a number of variational theorems in particular branches of continuum mechanics.

7.8 Problems

1. For an incompressible Newtonian viscous fluid in which Fourier's law of heat conduction is satisfied, T_{ij}, q_i and e are given by

$$T_{ij} = -p\delta_{ij} + 2\mu D_{ij}, \qquad q_i = -\kappa\, \partial T/\partial x_i, \qquad e = CT$$

where μ, κ and C are constants and T is the temperature. Deduce that in this case the energy equation (7.32) can be expressed in the form

$$\rho C\left(\frac{\partial T}{\partial t} + v_i \frac{\partial T}{\partial x_i}\right) = 2\mu D_{ij} D_{ij} + \kappa\, \nabla^2 T$$

2. A *singular surface* is a surface across which the stress, velocity and density may be discontinuous. By considering a thin cylindrical region which encloses part of a singular surface, show that in a body at rest in equilibrium, $\boldsymbol{t}^{(n)}$ is continuous across a stationary singular surface, where $\boldsymbol{n}$ is the normal to the singular surface.

3. Suppose a singular surface propagates through a body with speed V relative to the body, in the direction of the normal to the surface. Prove that the quantities ρV and $\rho V \boldsymbol{v} + \boldsymbol{t}^{(n)}$ are continuous across the singular surface.

4. A singular surface propagates in the direction of a unit vector $\boldsymbol{n}$ with speed v relative to fixed coordinates. Show that if $\boldsymbol{u}$ is continuous across the singular surface, then $v_i + v n_j\, \partial u_i/\partial x_j$ is also continuous across the singular surface.

8 Linear constitutive equations

8.1 Constitutive equations and ideal materials

The results given so far in this book apply equally to all materials. In themselves they are insufficient to describe the mechanical behaviour of any particular material.

To complete the specification of the mechanical properties of a material we require additional equations, which are called *constitutive equations*. These are equations which are particular to individual materials, or classes of materials, and they serve to distinguish one material from another. The mechanical constitutive equation of a material specifies the dependence of the stress in a body on kinematic variables such as a strain tensor or the rate-of-deformation tensor. Normally thermodynamic variables, especially temperature, will also be involved, but we shall make only brief references to these. Constitutive equations are also required in other branches of continuum physics, such as continuum thermodynamics and continuum electrodynamics, but these problems are outside the scope of this book, and we shall only discuss constitutive equations for the stress.

The mechanical behaviour of real materials is very diverse and complex and it would be impossible, even if it were desirable, to formulate equations which are capable of determining the stress in a body under all circumstances. Rather, we seek to establish equations which describe the most important features of the behaviour of a material in a given situation. Such equations can be regarded as defining *ideal materials*. It is unlikely that any real material will conform exactly to any such mathematical model, but if the ideal material is well chosen its behaviour may give an excellent approximation to that of the real material which it models. The model should be selected with the application as well as the material in mind, and the same real material may be represented by different ideal materials in different circumstances. For example, the theory of incompressible viscous fluids gives an excellent description of the behaviour of water flowing through pipes, but is useless for the study of the propagation of sound waves through water, because for sound-wave propagation a model which takes into account the compressibility of water is essential.

Historically, the constitutive equations which define the classical ideal materials (linear elastic solids, Newtonian viscous fluids, etc.) have been developed separately. In applications of these theories this separation is natural. However, at the formulative stage there are advantages in a unified approach which clarifies relations between the different special theories. Also it is possible to formulate some general principles which should be followed in the construction of constitutive equations.

A first requirement which any constitutive equation must satisfy is that of dimensional homogeneity: the dimensions of all terms in a constitutive equation must be the same. Since a constitutive equation always includes constants or functions which characterize the material under consideration, and these quantities have dimensions, the dimensional homogeneity requirement is usually not difficult to satisfy.

Constitutive equations should not depend on the choice of the coordinate system (although they may be expressed in terms of components relative to any selected coordinate system). They therefore take the form of relations between scalars, vectors and tensors.

An important restriction on mechanical constitutive equations is the requirement that the stress response of a body to a deformation is not affected by rigid-body motions, so that the stress in a body depends only on the change of shape of the body and is not affected (except for the change in orientation of the stress field relative to fixed axes) by a superposed motion in which the body moves as a whole. To formalize this requirement we specify that if a body undergoes two time-dependent motions, which differ from each other by a time-dependent rigid-body motion, then the same stress results from each of these motions. This is essentially equivalent to saying that constitutive equations are invariant under translations and rotations of the frame of reference; two observers, even if they are in relative motion, will observe the same stress in a given body.

Materials are usually regarded as either solids or fluids, and fluids are subdivided into liquids and gases. We do not attempt a precise definition of this classification; the dividing lines are not always clear and there are materials which possess both solid-like and fluid-like properties. The characteristic property of a fluid is that it cannot support a shearing stress indefinitely, so that if a shearing stress is applied to a body of fluid and maintained, the fluid will flow and continue to do so as long as the stress remains. A solid, on the other hand, can be in equilibrium under a shear stress. Some solids possess a natural configuration which they adopt in a stress-free state and to which they eventually return if a stress is imposed and then removed; if a natural configuration exists it is usually convenient, though not essential, to adopt it as the reference configuration. Fluids have no natural configuration and, given sufficient time, will adapt to the shape of any container in which they are placed.

8.2 Material symmetry

Most materials possess some form of *material symmetry*. The commonest case is that in which the material is *isotropic*; an isotropic material possesses no preferred direction and its properties are the same in all directions. It is impossible to detect the orientation in space of a sphere of isotropic material by performing an experiment on it. Many real materials are isotropic or nearly so; these include common fluids like air and water, metals in their usual polycrystalline form, concrete, sand in bulk, and so on. Other common materials have strong directional properties; an example is wood, whose properties along its grain are quite different from the properties across the grain. Single crystals of crystalline materials have directional properties which arise because their atoms are arranged in regular patterns, and this gives rise to the various classes of *crystal symmetry*. A material which possesses a single preferred direction at every point is said to be *transversely isotropic*. An example of such a material is a composite material which consists of a matrix reinforced by fibres arranged in parallel straight lines. Over length scales which are large compared to the fibre diameters and spacings, such a material may be regarded as macroscopically homogeneous, and the fibres introduce a preferred direction which is a characteristic of the composite material.

We consider material symmetries of two types; rotational and reflectional.

Rotational symmetry. Suppose a spherical volume element undergoes the homogeneous deformation illustrated in Fig. 8.1. A typical particle initially at P_0 moves to P_1 and the deformation is described by the equations

$$\boldsymbol{x} = \boldsymbol{F} \cdot \boldsymbol{X} \tag{8.1}$$

where, since the deformation is homogeneous, the components F_{iR} of $\boldsymbol{F}$ depend only on t.

Now suppose that the element undergoes a second deformation, which is similar to the first except that the entire deformation field (but not the body) is rotated through an angle α about an axis $\boldsymbol{n}$. Thus if $\boldsymbol{Q}$ is the tensor defined by (6.11), the particle which is initially at $\boldsymbol{Q} \cdot \boldsymbol{X}$ moves in the second deformation to the point $\boldsymbol{Q} \cdot \boldsymbol{x}$, where

$$\boldsymbol{Q} \cdot \boldsymbol{x} = \boldsymbol{F} \cdot \boldsymbol{Q} \cdot \boldsymbol{X} \tag{8.2}$$

The second deformation is illustrated for the case in which $\boldsymbol{n} = \boldsymbol{e}_3$ in Fig. 8.1(c); in it the particle initially at Q_0 moves to Q_2, where

$$\angle P_0 O Q_0 = \angle P_1 O Q_2 = \alpha$$

The deformed sphere has the same shape in the two configurations, but the second is not derived from the first by a rigid rotation. Although the two deformations (8.1) and (8.2) are related, they are distinct, and in the absence of appropriate material symmetry they will give rise to different stress responses. For example, the forces which accompany a given extension in the direction OP_0 will be different from those associated with the same extension in the direction OQ_0. However, for a given material it may happen that for certain rotations the result of rotating the deformation field through the rotation defined by $\boldsymbol{Q}$ is to

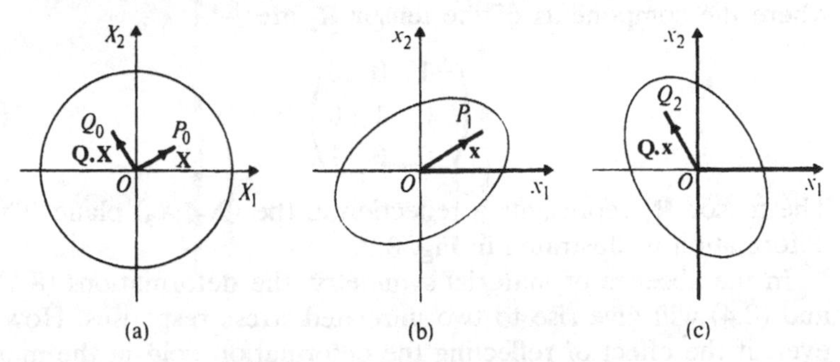

Figure 8.1 Rotational symmetry

produce the same rotation of the stress field. In this case, if the deformation (8.1) gives rise to a stress tensor $\boldsymbol{T}$, then the deformation (8.2) gives rise to a stress tensor $\boldsymbol{Q}^{\mathrm{T}} \cdot \boldsymbol{T} \cdot \boldsymbol{Q}$. We then say that the material has *material symmetry* (relative to the specified reference configuration) for the rotation determined by $\boldsymbol{Q}$.

As a simple example, the tensor $\boldsymbol{Q}$ with components Q_{iR}, where

$$(Q_{iR}) = \begin{pmatrix} 0 & 1 & 0 \\ -1 & 0 & 0 \\ 0 & 0 & 1 \end{pmatrix}$$

represents an anti-clockwise rotation of magnitude $\frac{1}{2}\pi$ about the X_3-axis. If the material has rotational symmetry for this rotation, then the force $\boldsymbol{p}_1$ required to produce a given extension in the X_1 direction has the same magnitude as the force $\boldsymbol{p}_2$ required to produce the same extension in the X_2 direction.

Reflectional symmetry. Now consider a further homogeneous deformation of the spherical volume element which is the mirror image of the deformation (8.1), in some plane which for definiteness we take to be the plane $X_1 = 0$. This deformation is defined by

$$\begin{pmatrix} -x_1 \\ x_2 \\ x_3 \end{pmatrix} = \begin{pmatrix} F_{11} & F_{12} & F_{13} \\ F_{21} & F_{22} & F_{23} \\ F_{31} & F_{32} & F_{33} \end{pmatrix} \begin{pmatrix} -X_1 \\ X_2 \\ X_3 \end{pmatrix} \tag{8.3}$$

or

$$\boldsymbol{R}_1 \cdot \boldsymbol{x} = \boldsymbol{F} \cdot \boldsymbol{R}_1 \cdot \boldsymbol{X} \tag{8.4}$$

where the components of the tensor $\boldsymbol{R}_1$ are

$$\begin{pmatrix} -1 & 0 & 0 \\ 0 & 1 & 0 \\ 0 & 0 & 1 \end{pmatrix} \tag{8.5}$$

The tensor $\boldsymbol{R}_1$ represents a reflection in the (X_2, X_3) plane. The deformation is illustrated in Fig. 8.2.

In the absence of material symmetry, the deformations (8.1) and (8.4) will give rise to two unrelated stress responses. However, if the effect of reflecting the deformation field in the manner described is to reverse the sign of the shear stress on the plane

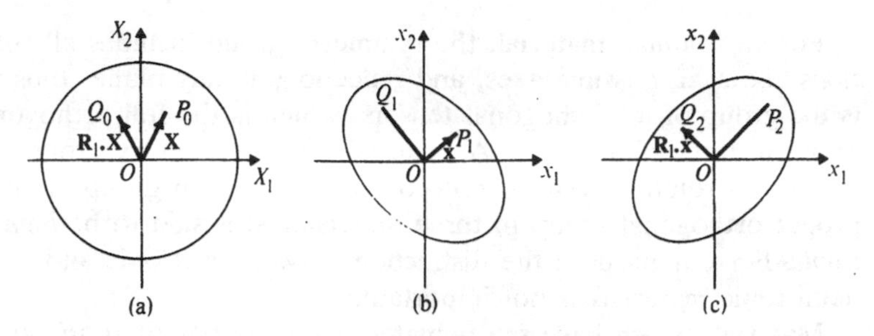

Figure 8.2 Reflectional symmetry

$x_1 = 0$, we say that the material has reflectional symmetry with respect to this plane, relative to the chosen reference configuration. If the material has this symmetry and the deformation (8.1) gives rise to the stress $\boldsymbol{T}$, then the deformation (8.4) gives rise to the stress $\boldsymbol{R}_1^{\mathrm{T}} \cdot \boldsymbol{T} \cdot \boldsymbol{R}_1$ (the transposition of $\boldsymbol{R}_1$ in the first factor is redundant, because $\boldsymbol{R}_1$ is symmetric, but is introduced for consistency with the corresponding result for rotational symmetries).

More generally, a reflection in the plane through O normal to a unit vector $\boldsymbol{n}$ is defined by a tensor $\boldsymbol{R}$ with components R_{ij}, where

$$\boldsymbol{R} = \boldsymbol{I} - 2\boldsymbol{n} \otimes \boldsymbol{n}, \qquad R_{ij} = \delta_{ij} - 2n_i n_j$$

It is easily verified that $\boldsymbol{R}$ is a symmetric improper orthogonal tensor (that is an orthogonal tensor with determinant equal to -1). A material has reflectional symmetry for reflections in the planes normal to $\boldsymbol{n}$ if the deformation

$$\boldsymbol{R} \cdot \boldsymbol{x} = \boldsymbol{F} \cdot \boldsymbol{R} \cdot \boldsymbol{X} \tag{8.6}$$

gives rise to the stress $\boldsymbol{R}^{\mathrm{T}} \cdot \boldsymbol{T} \cdot \boldsymbol{R}$ when the deformation (8.1) gives rise to the stress $\boldsymbol{T}$.

Reflectional symmetry with respect to planes normal to the X_1-axis means that the tangential force required to produce a simple shear in (say) the positive X_2 direction on the planes $X_1 =$ constant, is equal in magnitude but opposite in direction to that required to produce a shear of the same magnitude in the negative X_2 direction on the same planes.

Symmetry groups. The set of tensors, such as the rotation tensors $\boldsymbol{Q}$ and the reflection tensors $\boldsymbol{R}$, which define the symmetry properties of a material, form a group (in the technical algebraic sense of the term) which is called the *symmetry group* of the material.

For an *isotropic* material, the symmetry group includes all rotations about all possible axes, and reflections in any plane; thus it is the group of all orthogonal tensors, which is the full orthogonal group in three dimensions. A material whose symmetry group consists of all rotations but no reflections (the rotation group or the proper orthogonal group in three dimensions) is said to be *hemitropic.* For our purpose the distinction between isotropic and hemitropic materials is not important.

Materials which have fewer material symmetries than an isotropic material are said to be *anisotropic.* The symmetry group for an anisotropic material is a subgroup of the full orthogonal group.

A material whose symmetry group includes all rotations about a specified axis is said to be *transversely isotropic* about that axis. Various reflectional symmetries may or may not be added; again the distinctions are not important here.

A material which has reflectional symmetry with respect to each of three mutually orthogonal planes is said to be *orthotropic.* To a good approximation, wood is an example of such a material.

The symmetry group for an orthotropic material is a finite group, composed of the unit tensor, three reflection tensors, and their inner products. Other finite subgroups of the full orthogonal group in three dimensions are symmetry groups for materials with various kinds of *crystal symmetry.* The rotations which occur in these symmetry groups are rotations through multiples of $\frac{1}{2}\pi$ and $\frac{2}{3}\pi$. Accounts of the crystallographic groups can be found in texts on crystallography.

For the most part we shall concentrate on isotropy, which is the simplest and most important case, and make only occasional references to anisotropic materials.

8.3 Linear elasticity

Many solid materials, and especially the common engineering materials such as metals, concrete, wood, etc., have the property that they only undergo very small changes of shape when they are subjected to the forces which they normally encounter. They also have a natural shape to which they will return if forces are applied to them and then removed (provided that the forces are not too large). The theory of linear elasticity provides an excellent model of the mechanical behaviour of such materials.

We define a linear elastic solid to be a material for which the internal energy $\rho_0 e$ per unit volume in the reference configuration has the following properties:

(a) $\rho_0 e$ is a function only of the components E_{ij} of the infinitesimal strain tensor and is, or may be adequately approximated by, a quadratic function of these components;
(b) if K is the kinetic energy (7.27) and E is the internal energy (7.28) in any region $\mathcal{R}$, then the material time derivative of $K+E$ is equal to the rate at which mechanical work is done by the surface and body forces acting on $\mathcal{R}$.

It is conventional to denote $\rho_0 e$ by W, and to call W the strain–energy function. Thus (a) states that W has the form

$$W = \tfrac{1}{2} C_{ijkl} E_{ij} E_{kl} \tag{8.7}$$

where C_{ijkl} are constants. Property (b) is a restatement of the law of conservation of energy (Section 7.6) with heat flux assumed to be absent, or neglected. Properties (a) and (b) together state that all the mechanical work done on $\mathcal{R}$ either creates kinetic energy, or is stored as potential energy (which is called the *strain energy*) which depends only on the deformation. The system is conservative; in a closed cycle of deformation the strain energy is stored and then released so that no net work is done on the body.

The more general case in which W is allowed to depend also on temperature or entropy, and in which heat flux is permitted, leads to the theory of *linear thermoelasticity*. We shall not develop this theory.

It should be noted at the outset that a constitutive equation based on (8.7) will necessarily fail to satisfy one of the requirements stated in Section 8.1 for, as was shown in Section 6.6, the components E_{ij} do not remain constant in a finite rotation, and so W as defined by (8.7) must change when a body rotates without change of shape. This is not reasonable physically. However, if attention is restricted to motions in which the rotation is small, then the change in E_{ij} is of second order in the rotation components. The theory of linear elasticity is essentially an approximate theory which is valid for values of E_{ij} and Ω_{ij} which are small compared to one. The theory is nevertheless very useful because the approximation is an excellent one in many applications. It is consistent with the approximation involved in adopting (8.7) to neglect E_{ij} compared to one, and this will be done whenever it is convenient to do so.

Suppose we change from a coordinate system with base vectors $\boldsymbol{e}_i$ to a new coordinate system with base vectors $\bar{\boldsymbol{e}}_i$, such that

$$\bar{\boldsymbol{e}}_i = M_{ij}\boldsymbol{e}_j$$

and (M_{ij}) is an orthogonal matrix. Then the infinitesimal strain components E_{ij} and $\bar{E}_{ij}$ in the old and new systems are related by the usual tensor transformation rule

$$\bar{E}_{rs} = M_{ri}M_{sj}E_{ij}, \qquad E_{ij} = M_{ri}M_{sj}\bar{E}_{rs} \tag{8.8}$$

The strain energy W can also be expressed as a quadratic function of the components $\bar{E}_{ij}$, as

$$W = \tfrac{1}{2}\bar{C}_{ijkl}\bar{E}_{ij}\bar{E}_{kl} \tag{8.9}$$

However W is a scalar, which is not affected by a change of coordinate system, and so the expressions (8.7) and (8.9) are the same. Hence, using (8.8),

$$\bar{C}_{pqrs}\bar{E}_{pq}\bar{E}_{rs} = C_{ijkl}E_{ij}E_{kl} = C_{ijkl}M_{pi}M_{qj}M_{rk}M_{sl}\bar{E}_{pq}\bar{E}_{rs}$$

This is an identity for all values of $\bar{E}_{ij}$, and so

$$\bar{C}_{pqrs} = M_{pi}M_{qj}M_{rk}M_{sl}C_{ijkl}$$

Hence C_{ijkl} are components of a fourth-order tensor.

The $3^4 = 81$ constants C_{ijkl} are called *elastic constants.* They have the dimensions of stress and their values characterize particular linear elastic materials. The elastic constants are not all independent. By interchanging the dummy indices i and j in (8.7), we obtain

$$W = \tfrac{1}{2}C_{jikl}E_{ji}E_{kl}$$

However, $E_{ij} = E_{ji}$, and so

$$W = \tfrac{1}{2}C_{jikl}E_{ij}E_{kl} = \tfrac{1}{2}\{\tfrac{1}{2}(C_{ijkl} + C_{jikl})\}E_{ij}E_{kl}$$

Thus C_{ijkl} may be replaced by $\frac{1}{2}(C_{ijkl} + C_{jikl})$, which is symmetric with respect to interchanges of i and j. Hence, without loss of generality, C_{ijkl} may be assumed to be symmetric with respect to interchanges of its first two indices. Similarly, C_{ijkl} may be assumed to be symmetric with respect to interchanges of its third and fourth indices. Thus

$$C_{ijkl} = C_{jikl} = C_{ijlk} \qquad (i, j, k, l = 1, 2, 3) \tag{8.10}$$

The symmetries (8.10) reduce the number of independent elastic constants to 36. Furthermore, by simultaneously interchanging the indices i and k and the indices j and l, there follows

$$W = \tfrac{1}{2} C_{klij} E_{ij} E_{kl} = \tfrac{1}{2}\{\tfrac{1}{2}(C_{ijkl} + C_{klij})\} E_{ij} E_{kl}$$

Hence no generality is lost by assuming that C_{ijkl} also has the index symmetries

$$C_{ijkl} = C_{klij} \tag{8.11}$$

The symmetries (8.11) further reduce the number of independent elastic constants to 21.

A further requirement on W is that the stored elastic energy must be positive, so that (8.7) is a positive definite quadratic form in the E_{ij}.

Any material symmetry further reduces the number of independent elastic constants. We return to this point below.

So far, property (b) of linear elastic solids has not been employed. From (7.31), with e replaced by W/ρ_0, and the heat flux terms neglected, we have

$$T_{ij} \frac{\partial v_i}{\partial x_j} = \frac{\rho}{\rho_0} \frac{DW}{Dt} \tag{8.12}$$

Since, by (7.7) and (7.8), $\rho/\rho_0 = 1 + O(E_{ij})$, to the order of approximation used in small-deformation theory we may replace ρ by ρ_0, and write

$$T_{ij} \frac{\partial v_i}{\partial x_j} = \frac{DW}{Dt}$$

It was shown in Section 7.6 that $T_{ij}\, \partial v_i/\partial x_j = T_{ij} D_{ij}$, and so

$$T_{ij} D_{ij} = \frac{DW}{Dt} \tag{8.13}$$

Now, since W depends only on E_{ij}, (8.13) gives

$$T_{ij} D_{ij} = \frac{\partial W}{\partial E_{ij}} \frac{DE_{ij}}{Dt}$$

and (6.77) then gives, to the required order of approximation,

$$T_{ij} D_{ij} = \frac{\partial W}{\partial E_{ij}} D_{ij}$$

This is an identity which holds for all values of D_{ij}, and so

$$T_{ij}=\frac{\partial W}{\partial E_{ij}}$$

However, from (8.7) and (8.11),

$$\begin{aligned}\frac{\partial W}{\partial E_{ij}}&=\frac{1}{2}\frac{\partial}{\partial E_{ij}}(C_{pqrs}E_{pq}E_{rs})\\&=\tfrac{1}{2}C_{pqrs}(\delta_{ip}\delta_{jq}E_{rs}+\delta_{ir}\delta_{js}E_{pq})\\&=\tfrac{1}{2}(C_{ijrs}E_{rs}+C_{pqij}E_{pq})\\&=C_{ijrs}E_{rs}\end{aligned}$$

Hence

$$T_{ij}=C_{ijrs}E_{rs} \tag{8.14}$$

and this is the constitutive equation for a linear elastic solid. It is evident that the stress components are linear functions of the infinitesimal strain components.

An alternative formulation of linear elasticity theory is based on the assumption that the stress components T_{ij} are (or can adequately be approximated by) linear functions of the infinitesimal strain components E_{ij}, so that (8.14) is taken as the starting point rather than as a consequence of (8.7). In such a formulation there is no loss of generality in giving C_{ijkl} the index symmetries (8.10), but (8.11) does not obtain unless further assumptions are made. A material with constitutive equation (8.14) but lacking the index symmetry (8.11) has the unrealistic property that work can be extracted from it in a closed cycle of deformation. We therefore prefer to base the theory on (8.7), from which (8.11) follows automatically.

The number of independent elastic constants is further reduced if the material possesses any material symmetry. Suppose for example that the material has the reflectional symmetry with respect to the (X_2, X_3) planes which is associated with the tensor $\boldsymbol{R}_1$ which is defined by (8.5). Since $E_{ij}=\frac{1}{2}(F_{ij}+F_{ji})-\delta_{ij}$, it is easily seen that the effect of replacing the deformation (8.1) by the deformation (8.3) is to replace E_{12} by $-E_{12}$, and E_{13} by $-E_{13}$, while leaving the other components E_{ij} unaltered. However, if $\boldsymbol{R}_1$ belongs to the symmetry group, W must be unchanged by this substitution. Hence, if the material has this symmetry, then

$$W(E_{11},E_{22},E_{33},E_{23},E_{31},E_{12})=W(E_{11},E_{22},E_{33},E_{23},-E_{31},-E_{12}) \tag{8.15}$$

and this relation must hold identically for all E_{ij}. By writing (8.7) in full with the above two sets of arguments, or by considering special cases, it follows from (8.7) and (8.15) that

$$C_{1112}=C_{1113}=C_{1222}=C_{1223}=C_{1233}=C_{1322}=C_{1323}=C_{1333}=0$$

Other material symmetries impose further restrictions on the elastic constants. The various possibilities are described in texts on linear elasticity. We omit the details and proceed to the case of isotropic materials.

The symmetry group for isotropic materials includes all proper orthogonal tensors $\boldsymbol{Q}$. Suppose, as before, that E_{ij} are the components of infinitesimal strain which correspond to the deformation (8.1). Then the corresponding stress components T_{ij} are given by (8.14). The infinitesimal strain components which correspond to the deformation (8.2) are

$$\bar{E}_{pq}=Q_{kp}(\tfrac{1}{2}F_{kl}+\tfrac{1}{2}F_{lk}-\delta_{kl})Q_{lq}=Q_{kp}Q_{lq}E_{kl} \tag{8.16}$$

and the associated stress components are

$$\bar{T}_{rs}=C_{rspq}\bar{E}_{pq} \tag{8.17}$$

Now if $\boldsymbol{Q}$ belongs to the symmetry group, then

$$T_{ij}=Q_{ir}Q_{js}\bar{T}_{rs} \tag{8.18}$$

and hence, from (8.16), (8.17) and (8.18),

$$T_{ij}=Q_{ir}Q_{js}Q_{kp}Q_{lq}C_{rspq}E_{kl} \tag{8.19}$$

It follows, by comparing (8.14) and (8.19), that

$$C_{ijkl}=Q_{ir}Q_{js}Q_{kp}Q_{lq}C_{rspq} \tag{8.20}$$

and, if the material is isotropic, this must hold for all orthogonal tensors $\boldsymbol{Q}$. However, (8.20) then becomes a statement that C_{ijkl} are components of a fourth-order *isotropic* tensor (Section 3.5). The most general fourth-order isotropic tensor is given by (3.37). Hence C_{ijkl} take the form

$$C_{ijkl}=\lambda\delta_{ij}\delta_{kl}+\mu\delta_{ik}\delta_{jl}+\nu\delta_{il}\delta_{jk} \tag{8.21}$$

and the constitutive equation (8.14) becomes

$$T_{ij}=\lambda\delta_{ij}E_{kk}+\mu E_{ij}+\nu E_{ji}$$

Since $E_{ij}=E_{ji}$, no generality is lost by setting $\nu=\mu$, so that

$$T_{ij}=\lambda\delta_{ij}E_{kk}+2\mu E_{ij} \tag{8.22}$$

or, equivalently, in tensor notation

$$\boldsymbol{T} = \lambda \boldsymbol{I} \operatorname{tr} \boldsymbol{E} + 2\mu \boldsymbol{E}$$

Equation (8.22) is the constitutive equation for an isotropic linear elastic solid; such a material is characterized by the two elastic constants λ and μ.

We observe that the form (8.21) possesses the index symmetry $C_{ijkl} = C_{klij}$. Thus for an *isotropic* material we arrive at (8.22) regardless of whether we adopt (8.7) or (8.14) as the starting point.

8.4 Newtonian viscous fluids

In experiments on water, air and many other fluids, it is observed that in a simple shearing flow (Section 6.10) the shearing stress on the shear planes is proportional to the shear rate s, to an extremely good approximation and over a very wide range of shear rates. This behaviour is characteristic of a *Newtonian viscous fluid* or a *linear viscous fluid.* This model of fluid behaviour describes the mechanical properties of many fluids, including the commonest fluids, air and water, very well indeed.

We consider fluids with constitutive equations of the form

$$T_{ij} = -p(\rho, \theta)\delta_{ij} + B_{ijkl}(\rho, \theta)D_{kl} \qquad (8.23)$$

where θ is the temperature. In a fluid at rest, $D_{kl} = 0$, and (8.23) reduces to

$$T_{ij} = -p(\rho, \theta)\delta_{ij} \qquad (8.24)$$

which is the constitutive equation employed in hydrostatics, with $p(\rho, \theta)$ representing the hydrostatic pressure. Thus (8.23) specifies that in a fluid in motion the additional stress over the hydrostatic pressure is linear in the components of the rate of deformation tensor.

If the fluid is isotropic, then arguments similar to those used in Section 8.3 to reduce (8.14) to (8.22) lead to the conclusion that B_{ijkl} are (like C_{ijkl} for an isotropic linear elastic solid) the components of a fourth-order isotropic tensor, and then (8.23) takes the form

$$T_{ij} = \{-p(\rho, \theta) + \lambda(\rho, \theta)D_{kk}\}\delta_{ij} + 2\mu(\rho, \theta)D_{ij} \qquad (8.25)$$

or, equivalently

$$\boldsymbol{T} = \{-p(\rho, \theta) + \lambda(\rho, \theta) \operatorname{tr} \boldsymbol{D}\}\boldsymbol{I} + 2\mu(\rho, \theta)\boldsymbol{D}$$

Here the viscosity coefficients $\lambda(\rho, \theta)$ and $\mu(\rho, \theta)$ are of course, not the same as the elastic constants λ and μ which were introduced in Section 8.3. A particular linear viscous fluid is characterized by the two coefficients λ and μ.

It was shown in Section 6.9 that $D_{ij}=0$ in a rigid-body motion and that the superposition of a rigid-body motion on a given motion does not change the value of D_{ij}. Hence the right-hand side of (8.25) is not affected by a superimposed rigid-body motion. Therefore the constitutive equation (8.25) has the required property of being independent of superimposed rigid-body motions. This is in contrast to the constitutive equation of linear elasticity theory, which, it was emphasized in Section 8.3, is necessarily an approximate theory and is valid only for small rotations and deformations. Equation (8.25) is a possible exact constitutive equation for a viscous fluid. In practice, it is found that (8.25) serves extremely well to describe the mechanical behaviour of many fluids.

In fluid mechanics texts it is usual to assume, as we have done here, that the fluid is isotropic. In fact it can be shown that isotropy is a consequence of (8.23) and the requirement that the stress is not affected by rigid-body motions, and so isotropy need not be introduced as a separate assumption. We shall demonstrate this, in a more general context, in Section 10.3. It does not follow that all fluids are necessarily isotropic. Fluids with anisotropic properties do exist, but they require more general constitutive equations than (8.23) for their description.

Several special cases of (8.25) are of interest. If the stress is a hydrostatic pressure (see Section 5.9) then

$$T_{ij}=\tfrac{1}{3}T_{kk}\delta_{ij}=\{-p(\rho, \theta)+(\lambda+\tfrac{2}{3}\mu)D_{kk}\}\delta_{ij}$$

It is often assumed that in such a state of pure hydrostatic stress, the stress depends only on ρ and θ and not on the dilatation rate D_{kk}. If this is the case then $\lambda+\tfrac{2}{3}\mu=0$, and this relation is often adopted.

If the material is *inviscid*, then $\lambda=0$ and $\mu=0$, and the constitutive equation reduces to (8.24). The stress in an inviscid fluid is always hydrostatic.

If the fluid is *incompressible*, then ρ is constant and $D_{kk}=0$. Incompressibility is a kinematic constraint which gives rise to a reaction stress. The reaction to incompressibility is an arbitrary hydrostatic pressure which can be superimposed on the stress field without causing any deformation; this pressure does no work in any deformation which satisfies the incompressibility constraint. Such a hydrostatic pressure is not determined by constitutive equations but can only be found through the equations of motion or of equilibrium, and the boundary conditions. Thus for an incompressible viscous fluid, (8.25) reduces to

$$T_{ij} = -p\delta_{ij} + 2\mu(\theta)D_{ij}, \qquad \text{or} \qquad \boldsymbol{T} = -p\boldsymbol{I} + 2\mu(\theta)\boldsymbol{D} \tag{8.26}$$

where p is arbitrary, μ depends only on θ, and the term λD_{kk} has been absorbed into the arbitrary function p. We note that in the limit as the material becomes incompressible, $D_{kk} \to 0$ and $\lambda \to \infty$ in such a way that λD_{kk} tends to a finite limit.

If the fluid is both inviscid and incompressible (such a fluid is called an *ideal fluid*) then

$$T_{ij} = -p\delta_{ij}, \qquad \text{or} \qquad \boldsymbol{T} = -p\boldsymbol{I} \tag{8.27}$$

where p is arbitrary in the sense that it is not determined by a constitutive equation.

8.5 Linear viscoelasticity

Many materials (especially materials which are usually described as 'plastics') possess both some of the characteristics of elastic solids and some of the characteristics of viscous fluids. Such materials are termed *viscoelastic.* The phenomenon of viscoelasticity is illustrated by *creep* and *stress-relaxation* experiments. For simplicity, consider the case of simple tension. Suppose a tension F_0 is rapidly applied to an initially stress-free viscoelastic string at time $t = 0$, and then held, constant, as illustrated in Fig. 8.3(a). The

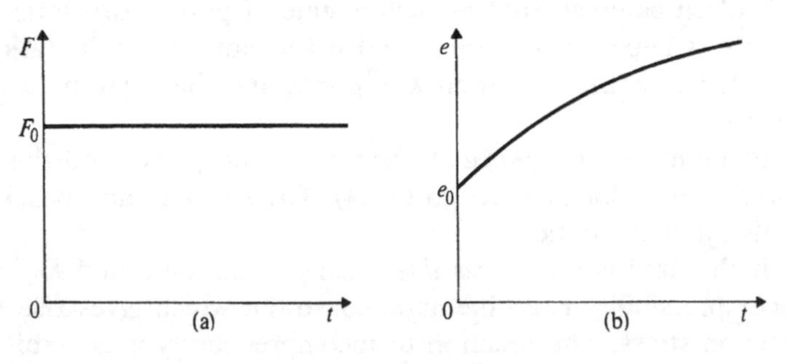

Figure 8.3 Creep curve

corresponding relation between the elongation e and time t may be of the form shown in Fig. 8.3(b), with an initial elongation e_0 (such as would occur in an elastic material) followed by an increasing elongation under the maintained load. This illustrates the phenomenon of creep. If the material is a viscoelastic *solid*, the elongation tends to a finite limit e_∞ as $t \to \infty$; if the material is a viscoelastic *fluid*, the elongation continues indefinitely.

Alternatively, suppose that at $t=0$ the string is given an elongation e_0 and held in this position (Fig. 8.4(a)). The resulting force response is shown in Fig. 8.4(b); the force rises instantaneously to F_0 at $t=0$ and then decays. This is stress relaxation. For a fluid, $F \to 0$ as $t \to \infty$; in a solid, F tends to a finite limit F_∞ as $t \to \infty$.

We consider here only infinitesimal deformations, so that the use of the infinitesimal strain tensor is appropriate. With the behaviour illustrated in Fig. 8.4 as motivation, we assume that an increment δE_{ij} in the strain components at time τ gives rise to increments δT_{ij} in the stress components at subsequent times t, the magnitude of these increments depending on the lapse of time since the strain increment was applied. Thus

$$\delta T_{ij}(t) = G_{ijkl}(t-\tau)\, \delta E_{kl}(\tau) \tag{8.28}$$

where we expect G_{ijkl} to be decreasing functions of $t-\tau$. The *superposition principle* is also assumed, according to which the total stress at time t is obtained by superimposing the effect at time t of all the strain increments at times $\tau < t$. Thus

$$T_{ij}(t) = \int_{-\infty}^{t} G_{ijkl}(t-\tau) \frac{\mathrm{d}E_{kl}(\tau)}{\mathrm{d}\tau}\, \mathrm{d}\tau \tag{8.29}$$

This is the constitutive equation for linear viscoelasticity. The functions G_{ijkl} are called *relaxation functions*. If the strain was

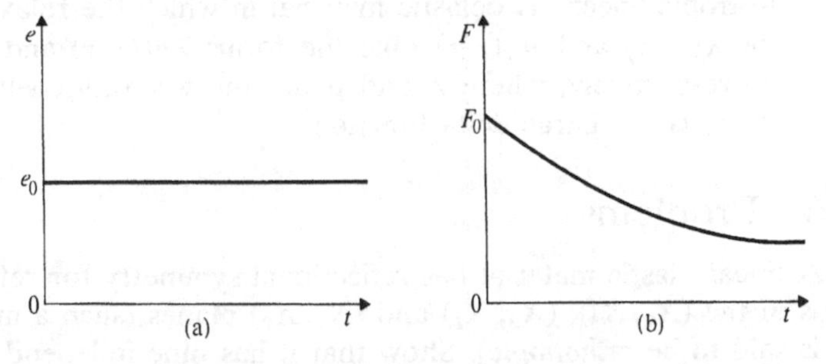

Figure 8.4 Stress-relaxation curve

zero in the remote past, so that $E_{kl} \to 0$ as $\tau \to -\infty$, (8.29) can be expressed in an alternative form by carrying out an integration by parts, as follows:

$$T_{ij}(t) = E_{kl}(t)G_{ijkl}(0) - \int_{-\infty}^{t} E_{kl} \frac{\mathrm{d}}{\mathrm{d}\tau}\{G_{ijkl}(t-\tau)\}\,\mathrm{d}\tau \qquad (8.30)$$

The stress-relaxation functions $G_{ijkl}(t-\tau)$ have the index symmetries $G_{ijkl} = G_{jikl} = G_{ijlk}$, but not the index symmetry $G_{ijkl} = G_{klij}$, unless this is introduced as a further assumption. If the material is isotropic, then G_{ijkl} are components of a fourth-order isotropic tensor, and, for example, (8.29) reduces to

$$T_{ij} = \delta_{ij} \int_{-\infty}^{t} \lambda(t-\tau) \frac{\mathrm{d}E_{kk}(\tau)}{\mathrm{d}\tau}\,\mathrm{d}\tau + 2\int_{-\infty}^{t} \mu(t-\tau) \frac{\mathrm{d}E_{ij}(\tau)}{\mathrm{d}\tau}\,\mathrm{d}\tau \qquad (8.31)$$

and only two relaxation functions $\lambda(t-\tau)$ and $\mu(t-\tau)$ are required to describe the material.

The inverse relation to (8.29) is

$$E_{ij}(\tau) = \int_{-\infty}^{t} J_{ijkl}(t-\tau) \frac{\mathrm{d}T_{kl}}{\mathrm{d}\tau}\,\mathrm{d}\tau \qquad (8.32)$$

The functions $J_{ijkl}(t-\tau)$ are known as *creep functions*; they have the same index symmetries as $G_{ijkl}(t-\tau)$ and are components of a fourth-order isotropic tensor in the case in which the material is isotropic.

Linear viscoelasticity has the same limitations as linear elasticity: it is necessarily an approximate theory which can only be applicable when the strain and rotation components are small.

In a sense, linear elasticity can be regarded as the limiting case of linear viscoelasticity in which the relaxation functions are independent of t; and a Newtonian viscous fluid as the limiting case of an isotropic linear viscoelastic material in which the relaxation functions $\lambda(t-\tau)$ and $\mu(t-\tau)$ take the forms $\lambda\delta(t-\tau)$ and $\mu\delta(t-\tau)$ respectively, where λ and μ are the viscosity coefficients and $\delta(t-\tau)$ is the Dirac delta function.

8.6 Problems

1. A linear elastic material has reflectional symmetry for reflections in the (X_2, X_3), (X_3, X_1) and (X_1, X_2) planes (such a material is said to be *orthotropic*). Show that it has nine independent elastic constants.

2. Show that a transversely isotropic linear elastic solid has five independent elastic constants, and find the form of W for a linear elastic solid which is transversely isotropic with respect to the X_3-axis.

3. From the constitutive equation (8.22) and the equation of motion (7.22), with $\boldsymbol{b}=\mathbf{0}$, derive Navier's equations for an isotropic linear elastic solid:

$$(\lambda+\mu)\frac{\partial^2 u_k}{\partial x_k\,\partial x_i}+\mu\frac{\partial^2 u_i}{\partial x_k^2}=\rho\frac{\partial^2 u_i}{\partial t^2}$$

4. In simple tension of an isotropic linear elastic solid. $T_{11}=EE_{11}$, $T_{22}=T_{33}=T_{23}=T_{31}=T_{12}=0$, and $E_{22}=E_{33}=-\nu E_{11}$, where E is Young's modulus and ν is Poisson's ratio. Prove that $E=\mu(3\lambda+2\mu)/(\lambda+\mu)$ and $\nu=\frac{1}{2}\lambda/(\lambda+\mu)$. Show that the constitutive equation (8.22) can be expressed in the form

$$\boldsymbol{E}=\frac{1}{E}\{(1+\nu)\boldsymbol{T}-\nu\boldsymbol{I}\operatorname{tr}\boldsymbol{T}\}$$

5. Prove that necessary and sufficient conditions for W to be positive definite for an isotropic linear elastic solid are $\mu>0$, $\lambda+\frac{2}{3}\mu>0$.

6. In plane stress or in plane strain, the equilibrium equations reduce to (5.42). Show that these equations are identically satisfied if the stress components are expressed in terms of Airy's stress function χ, as $T_{11}=\partial^2\chi/\partial x_2^2$, $T_{22}=\partial^2\chi/\partial x_1^2$, $T_{12}=-\partial^2\chi/\partial x_1\,\partial x_2$. Prove that in plane stress or plane strain of an isotropic linear elastic solid, χ satisfies the biharmonic equation

$$\nabla^4\chi=\left(\frac{\partial^2}{\partial x_1^2}+\frac{\partial^2}{\partial x_2^2}\right)^2\chi=0$$

7. From the constitutive equation (8.26) and the equations of motion (7.22) derive the Navier–Stokes equations for an incompressible Newtonian fluid

$$\rho b_i-\frac{\partial p}{\partial x_i}+\mu\frac{\partial^2 v_i}{\partial x_j\,\partial x_j}=\rho\left(\frac{\partial v_i}{\partial t}+v_j\frac{\partial v_i}{\partial x_j}\right)$$

8. A Voigt solid is a model viscoelastic material which in uniaxial tension has the stress–strain relation $\sigma=E_0(\varepsilon+t_0\dot{\varepsilon})$, where E_0 and t_0 are constants. Sketch the creep and stress-relaxation curves

for this material. Show that the relaxation function is $E_0\{1 + t_0\,\delta(t-\tau)\}$. Give a three-dimensional generalization of the above constitutive equation for an incompressible isotropic material.

9. A Maxwell fluid is a model viscoelastic material which in uniaxial tension has the stress–strain relation $E_1\dot{\varepsilon} = \dot{\sigma} + \sigma/t_1$. Sketch the creep and stress relaxation curves. Show that the stress relaxation function is $E_1 \exp\{-(t-\tau)/t_1\}$. Hence give a three-dimensional generalization for an isotropic incompressible material in the integral form (8.31).

9

Further analysis of finite deformation

9.1 Deformation of a surface element

The extension of a material line element in the deformation (6.1) was discussed in Section 6.2 and the change of volume of a material volume element was considered in Section 7.2. In some applications it is important to know how the area and orientation of a material surface element change in a deformation; this problem arises, for example, when specified forces are applied to the boundary of a deforming body.

Consider a triangular material surface element whose vertices P_0, Q_0 and R_0 in the reference configuration have position vectors $\boldsymbol{X}^{(0)}$, $\boldsymbol{X}^{(0)}+\delta\boldsymbol{X}^{(1)}$ and $\boldsymbol{X}^{(0)}+\delta\boldsymbol{X}^{(2)}$ respectively, as shown in Fig. 9.1. Let this triangle have area δS and unit normal vector $\boldsymbol{N}$.* Then by elementary vector algebra,

$$\boldsymbol{N}\,\delta S=\tfrac{1}{2}\delta\boldsymbol{X}^{(1)}\times\delta\boldsymbol{X}^{(2)},\qquad\text{or}\qquad N_R\,\delta S=\tfrac{1}{2}e_{RST}\,\delta X_S^{(1)}\,\delta X_T^{(2)}\tag{9.1}$$

Suppose that in the deformation (6.1) the particles initially at P_0, Q_0 and R_0 move to the positions P, Q and R, with respective position vectors $\boldsymbol{x}^{(0)}$, $\boldsymbol{x}^{(0)}+\delta\boldsymbol{x}^{(1)}$ and $\boldsymbol{x}^{(0)}+\delta\boldsymbol{x}^{(2)}$, and that the triangle $P_0Q_0R_0$ has area δs and unit normal $\boldsymbol{n}$. Then

$$\boldsymbol{n}\,\delta s=\tfrac{1}{2}\delta\boldsymbol{x}^{(1)}\times\delta\boldsymbol{x}^{(2)},\qquad\text{or}\qquad n_i\,\delta s=\tfrac{1}{2}e_{ijk}\,\delta x_j^{(1)}\,\delta_k^{(2)}\tag{9.2}$$

We now introduce (7.3) and the similar relation for $\delta x_i^{(2)}$ into (9.2), and so obtain

$$\begin{aligned}n_i\,\delta s&=\tfrac{1}{2}e_{ijk}\,\delta x_j^{(1)}\,\delta x_k^{(2)}\\&=\tfrac{1}{2}e_{ijk}\frac{\partial x_j}{\partial X_S}\frac{\partial x_k}{\partial X_T}\,\delta X_S^{(1)}\,\delta X_T^{(2)}\\&\quad+O(\delta X_R^{(\alpha)})^3\qquad(\alpha=1,2)\end{aligned}$$

* The use of $\boldsymbol{N}$ to denote a vector is another departure from the convention that vectors are denoted by lower-case letters.

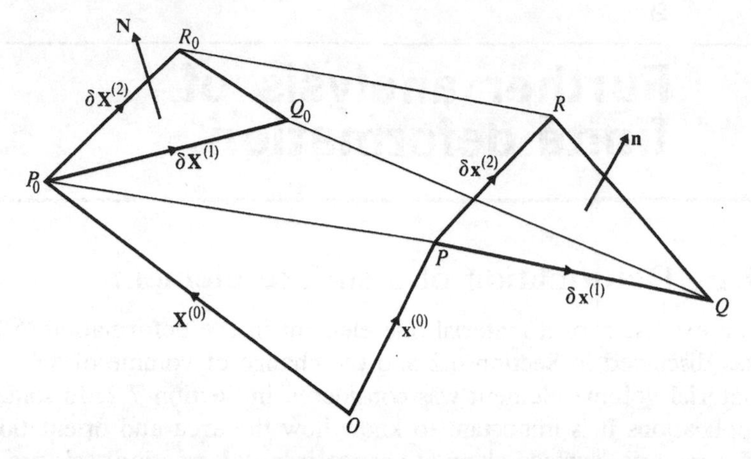

Figure 9.1 Deformation of a surface element

Next multiply both sides of this equation by $\partial x_i/\partial X_R$. This gives

$$n_i \frac{\partial x_i}{\partial X_R}\,\delta s = \tfrac{1}{2} e_{ijk} \frac{\partial x_i}{\partial X_R} \frac{\partial x_j}{\partial X_S} \frac{\partial x_k}{\partial X_T}\,\delta X_S^{(1)}\,\delta X_T^{(2)} + O(\delta X_R^{(\alpha)})^3$$

It then follows from (2.22) and (9.1) that

$$n_i \frac{\partial x_i}{\partial X_R}\,\delta s = \tfrac{1}{2} e_{RST} \frac{\partial(x_1, x_2, x_3)}{\partial(X_1, X_2, X_3)}\,\delta X_S^{(1)}\,\delta X_T^{(2)} + O(\delta X_R^{(\alpha)})^3$$

$$= \frac{\partial(x_1, x_2, x_3)}{\partial(X_1, X_2, X_3)} N_R\,\delta S + O(\delta X_R^{(\alpha)})^3 \quad (9.3)$$

In the limit as $\delta X^{(1)} \to 0$, and $\delta X^{(2)} \to 0$, (9.3) becomes

$$n_i \frac{\partial x_i}{\partial X_R} \frac{\mathrm{d}s}{\mathrm{d}S} = \frac{\partial(x_1, x_2, x_3)}{\partial(X_1, X_2, X_3)} N_R = (\det \boldsymbol{F}) N_R = \frac{\rho_0}{\rho} N_R \quad (9.4)$$

Since $\boldsymbol{N}$ is a unit vector, it follows from (9.4) that

$$1 = N_R N_R = (\det \boldsymbol{F})^{-2} n_i n_j \frac{\partial x_i}{\partial X_R} \frac{\partial x_j}{\partial X_R} \left(\frac{\mathrm{d}s}{\mathrm{d}S}\right)^2$$

$$= (\det \boldsymbol{F})^{-2} n_i n_j B_{ij} \left(\frac{\mathrm{d}s}{\mathrm{d}S}\right)^2 \quad (9.5)$$

and hence that

$$\left(\frac{\mathrm{d}s}{\mathrm{d}S}\right)^2 = \frac{(\det \boldsymbol{F})^2}{n_i n_j B_{ij}} \quad (9.6)$$

In tensor notation, (9.4) and (9.6) may be written as

$$\boldsymbol{N}\det\boldsymbol{F}=\boldsymbol{n}\cdot\boldsymbol{F}\frac{\mathrm{d}s}{\mathrm{d}S} \tag{9.7}$$

and

$$\left(\frac{\mathrm{d}s}{\mathrm{d}S}\right)^2=\frac{(\det\boldsymbol{F})^2}{\boldsymbol{n}\cdot\boldsymbol{B}\cdot\boldsymbol{n}} \tag{9.8}$$

Equations (9.6) or (9.8) determine the area ratio $\mathrm{d}s/\mathrm{d}S$ in terms of the deformation and the normal $\boldsymbol{n}$ in the deformed configuration. The initial normal $\boldsymbol{N}$ is then given by (9.4) or (9.7). The inverse relations to (9.7) and (9.8) are

$$\boldsymbol{n}\det\boldsymbol{F}^{-1}=\boldsymbol{N}\cdot\boldsymbol{F}^{-1}\frac{\mathrm{d}S}{\mathrm{d}s} \tag{9.9}$$

and

$$\left(\frac{\mathrm{d}S}{\mathrm{d}s}\right)^2=\frac{(\det\boldsymbol{F}^{-1})^2}{\boldsymbol{N}\cdot\boldsymbol{C}^{-1}\cdot\boldsymbol{N}} \tag{9.10}$$

9.2 Decomposition of a deformation

By the polar decomposition theorem (Sections 2.5, 3.6) the deformation-gradient tensor $\boldsymbol{F}$ may be expressed in the forms

$$\boldsymbol{F}=\boldsymbol{R}\cdot\boldsymbol{U}=\boldsymbol{V}\cdot\boldsymbol{R} \tag{9.11}$$

where $\boldsymbol{R}$ is an orthogonal tensor and $\boldsymbol{U}$ and $\boldsymbol{V}$ are symmetric positive definite tensors. Since $\det\boldsymbol{F}=\rho_0/\rho$, it can be assumed that $\det\boldsymbol{F}>0$, and then $\boldsymbol{R}$ is a proper orthogonal tensor. For a given tensor $\boldsymbol{F}$, the tensors $\boldsymbol{R}$, $\boldsymbol{U}$ and $\boldsymbol{V}$ are unique. It follows immediately from (9.11) that

$$\boldsymbol{U}=\boldsymbol{R}^{\mathrm{T}}\cdot\boldsymbol{V}\cdot\boldsymbol{R},\qquad \boldsymbol{V}=\boldsymbol{R}\cdot\boldsymbol{U}\cdot\boldsymbol{R}^{\mathrm{T}} \tag{9.12}$$

We consider first the case in which the motion is homogeneous, so that

$$\boldsymbol{x}=\boldsymbol{F}\cdot\boldsymbol{X} \tag{9.13}$$

where the components of $\boldsymbol{F}$ are constants. Suppose that the body undergoes two successive homogeneous motions, in which the particle which initially has position vector $\boldsymbol{X}$ moves first to the point with position vector $\hat{\boldsymbol{x}}$ and secondly to the point with position vector $\boldsymbol{x}$, where

$$\hat{\boldsymbol{x}} = \boldsymbol{U} \cdot \boldsymbol{X}, \qquad \boldsymbol{x} = \boldsymbol{R} \cdot \hat{\boldsymbol{x}} \tag{9.14}$$

Then from (9.11) and (9.14)

$$\boldsymbol{x} = \boldsymbol{R} \cdot \hat{\boldsymbol{x}} = \boldsymbol{R} \cdot \boldsymbol{U} \cdot \boldsymbol{X} = \boldsymbol{F} \cdot \boldsymbol{X}$$

and the two successive motions (9.14) are equivalent to the motion (9.13). Since $\boldsymbol{R}$ is orthogonal, the second equation of (9.14) describes a rotation of the body. The first equation of (9.14) describes a deformation which corresponds to the symmetric tensor $\boldsymbol{U}$. Thus the first equation of (9.11) shows that any homogeneous deformation can be decomposed into a deformation which corresponds to the symmetric tensor $\boldsymbol{U}$ followed by the rotation $\boldsymbol{R}$. Similarly, the second equation of (9.11) shows that, alternatively, any homogeneous deformation can be decomposed into the same rotation $\boldsymbol{R}$ followed by a deformation which corresponds to the symmetric tensor $\boldsymbol{V}$.

If the deformation is not homogeneous, (9.13) may be replaced by the relation

$$\mathrm{d}\boldsymbol{x} = \boldsymbol{F} \cdot \mathrm{d}\boldsymbol{X}$$

between the differentials $\mathrm{d}\boldsymbol{x}$ and $\mathrm{d}\boldsymbol{X}$. Then the decompositions (9.11) can still be made, but $\boldsymbol{R}$, $\boldsymbol{U}$ and $\boldsymbol{V}$ are now functions of position. In this case the decomposition is regarded as one into a *local* deformation $\boldsymbol{U}$ followed by a *local* rotation $\boldsymbol{R}$, or alternatively into a *local* rotation $\boldsymbol{R}$ followed by the local deformation $\boldsymbol{V}$.

The tensor $\boldsymbol{R}$ is called the *rotation tensor.* The tensors $\boldsymbol{U}$ and $\boldsymbol{V}$ are called the *right stretch* and the *left stretch tensors* respectively. The tensors $\boldsymbol{U}$ and $\boldsymbol{V}$ are closely related to the deformation tensors $\boldsymbol{C}$ and $\boldsymbol{B}$, for, from (6.27) and (9.11), and since $\boldsymbol{U}$ is symmetric, we have

$$\boldsymbol{C} = \boldsymbol{F}^{\mathrm{T}} \cdot \boldsymbol{F} = \boldsymbol{U} \cdot \boldsymbol{R}^{\mathrm{T}} \cdot \boldsymbol{R} \cdot \boldsymbol{U} = \boldsymbol{U}^2 \tag{9.15}$$

and from (6.33) and (9.11) we have

$$\boldsymbol{B} = \boldsymbol{F} \cdot \boldsymbol{F}^{\mathrm{T}} = \boldsymbol{V} \cdot \boldsymbol{R} \cdot \boldsymbol{R}^{\mathrm{T}} \cdot \boldsymbol{V} = \boldsymbol{V}^2 \tag{9.16}$$

Because $\boldsymbol{U}$ is symmetric and positive definite, (9.15) determines the components of $\boldsymbol{U}$ in terms of those of $\boldsymbol{C}$, and conversely. Therefore $\boldsymbol{U}$ and $\boldsymbol{C}$ are measures of the deformation which are equivalent to each other. $\boldsymbol{U}$ has the advantage of possessing the geometrical interpretation described in this section. However, for a given $\boldsymbol{F}$, the direct calculation of $\boldsymbol{U}$ from (9.11) is inconvenient, whereas the calculation of $\boldsymbol{C}$ from (6.27) is straightforward. Therefore in applications the use of $\boldsymbol{C}$ is usually to be preferred to that of $\boldsymbol{U}$. Similar comments apply to the tensors $\boldsymbol{B}$ and $\boldsymbol{V}$.

From (6.62) we have

$$\boldsymbol{F}=\boldsymbol{I}+\boldsymbol{E}+\boldsymbol{\Omega} \tag{9.17}$$

where $\boldsymbol{E}$ is symmetric and $\boldsymbol{\Omega}$ is anti-symmetric. In the case of small strains and rotations, we neglect squares and products of $\boldsymbol{E}$ and $\boldsymbol{\Omega}$. Then

$$\boldsymbol{U}^2=\boldsymbol{F}^{\mathrm{T}}\cdot\boldsymbol{F}=(\boldsymbol{I}+\boldsymbol{E}-\boldsymbol{\Omega})\cdot(\boldsymbol{I}+\boldsymbol{E}+\boldsymbol{\Omega})\simeq\boldsymbol{I}+2\boldsymbol{E}$$

and, to the same order of approximation,

$$\boldsymbol{U}\simeq\boldsymbol{I}+\boldsymbol{E} \tag{9.18}$$

In a similar way we find that $\boldsymbol{V}\simeq\boldsymbol{I}+\boldsymbol{E}$, so that both $\boldsymbol{U}-\boldsymbol{I}$ and $\boldsymbol{V}-\boldsymbol{I}$ reduce to the infinitesimal strain tensor in the case of small deformations. Also, from (9.18),

$$\boldsymbol{U}^{-1}\simeq\boldsymbol{I}-\boldsymbol{E} \tag{9.19}$$

and so from (9.11), (9.17) and (9.19)

$$\boldsymbol{R}=\boldsymbol{F}\cdot\boldsymbol{U}^{-1}\simeq(\boldsymbol{I}+\boldsymbol{E}+\boldsymbol{\Omega})\cdot(\boldsymbol{I}-\boldsymbol{E})\simeq\boldsymbol{I}+\boldsymbol{\Omega} \tag{9.20}$$

Thus $\boldsymbol{R}-\boldsymbol{I}$ reduces to the infinitesimal rotation tensor $\boldsymbol{\Omega}$ in the case of small rotations.

9.3 Principal stretches and principal axes of deformation

Suppose that $\boldsymbol{F}$ has been decomposed into the product $\boldsymbol{R}\cdot\boldsymbol{U}$, as in (9.11). The factor $\boldsymbol{R}$ represents a rotation. We now concentrate on the motion which corresponds to the symmetric tensor $\boldsymbol{U}$.

We recall the result (6.20) which gives the change of orientation of a material line element in a motion. For the motion $\boldsymbol{U}$, this result becomes

$$\boldsymbol{U}\cdot\mathbf{A}=\lambda\boldsymbol{a} \tag{9.21}$$

where $\boldsymbol{A}$ and $\boldsymbol{a}$ are unit vectors in the direction of the line element before and after the motion $\boldsymbol{U}$, and λ is the stretch of the element.

Suppose a particular line element, whose initial direction is given by $\boldsymbol{A}$, stretches but does not rotate during the motion. Then for this line element $\boldsymbol{A}$ is equal to $\boldsymbol{a}$, and (9.21) becomes

$$\boldsymbol{U} \cdot \boldsymbol{A} = \lambda \boldsymbol{A}$$

or

$$(\boldsymbol{U} - \lambda \boldsymbol{I}) \cdot \boldsymbol{A} = \boldsymbol{0} \tag{9.22}$$

Thus λ is a principal value of $\boldsymbol{U}$, and $\boldsymbol{A}$ is a principal direction of $\boldsymbol{U}$. Since $\boldsymbol{U}$ is symmetric and positive definite, its principal values are real and positive; we denote them by λ_1, λ_2 and λ_3, order them so that $\lambda_1 \geqslant \lambda_2 \geqslant \lambda_3$, and call them the *principal stretches.* Also since $\boldsymbol{U}$ is symmetric, it has a triad of orthogonal principal directions given by unit vectors $\mathbf{A}_1$, $\mathbf{A}_2$ and $\mathbf{A}_3$, which are uniquely determined if λ_1, λ_2 and λ_3 are distinct. These vectors determine the *principal axes* of $\boldsymbol{U}$.

If the coordinate axes are chosen to coincide with the principal axes of $\boldsymbol{U}$, then the matrix of the components of $\boldsymbol{U}$ takes the diagonal form

$$(U_{RS}) = \begin{pmatrix} \lambda_1 & 0 & 0 \\ 0 & \lambda_2 & 0 \\ 0 & 0 & \lambda_3 \end{pmatrix}$$

Hence, referred to these axes, the deformation $\boldsymbol{U}$ consists of extensions along the three coordinate directions, with no rotation of elements which lie along these axes. Therefore the motion which corresponds to $\boldsymbol{F} = \boldsymbol{R} \cdot \boldsymbol{U}$ consists of these three extensions of magnitudes λ_1, λ_2 and λ_3 along the three directions $\mathbf{A}_1$, $\mathbf{A}_2$ and $\boldsymbol{A}_3$ respectively, followed by the rotation $\boldsymbol{R}$.

In a similar way, the decomposition $\boldsymbol{F} = \mathbf{V} \cdot \boldsymbol{R}$ can be used to show that, alternatively, $\boldsymbol{F}$ can be regarded as a rotation $\boldsymbol{R}$ followed by three extensions, which are given by the principal values of $\boldsymbol{V}$, along the directions of the principal axes of $\boldsymbol{V}$. However, the principal values and principal axes of $\boldsymbol{U}$ and $\mathbf{V}$ are related. Since $\boldsymbol{R}^{\mathrm{T}} \cdot \boldsymbol{R} = \boldsymbol{I}$, it follows from (9.22) that

$$\boldsymbol{R} \cdot (\boldsymbol{U} - \lambda \boldsymbol{I}) \cdot \boldsymbol{R}^{\mathrm{T}} \cdot \boldsymbol{R} \cdot \mathbf{A} = \mathbf{0}$$

Since $\boldsymbol{R} \cdot \boldsymbol{I} \cdot \boldsymbol{R}^{\mathrm{T}} = \boldsymbol{I}$, this equation can be expressed as

$$(\boldsymbol{R} \cdot \boldsymbol{U} \cdot \boldsymbol{R}^{\mathrm{T}} - \lambda \boldsymbol{I}) \cdot \boldsymbol{R} \cdot \mathbf{A} = \mathbf{0}$$

and hence, from (9.12), as

$$(\boldsymbol{V}-\lambda\boldsymbol{I})\cdot\boldsymbol{R}\cdot\boldsymbol{A}=\boldsymbol{0} \tag{9.23}$$

Thus the principal stretches λ_1, λ_2 and λ_3 of $\boldsymbol{U}$ are also the principal values of $\boldsymbol{V}$, and if $\boldsymbol{A}_1$, $\boldsymbol{A}_2$ and $\boldsymbol{A}_3$ define the principal directions of $\boldsymbol{U}$, then $\boldsymbol{R}\cdot\boldsymbol{A}_1$, $\boldsymbol{R}\cdot\boldsymbol{A}_2$ and $\boldsymbol{R}\cdot\boldsymbol{A}_3$ define the principal directions of $\boldsymbol{V}$. The principal directions of $\boldsymbol{V}$ are obtained by rotating the principal directions of $\boldsymbol{U}$ through the rotation $\boldsymbol{R}$.

If the deformation is homogeneous, then $\boldsymbol{U}$, $\boldsymbol{V}$ and $\boldsymbol{R}$ are constant tensors, and the principal stretches and the principal directions are uniform throughout the body. In the general case of a non-homogeneous deformation, the principal stretches λ_1, λ_2 and λ_3, and the vectors $\boldsymbol{A}_1$, $\boldsymbol{A}_2$ and $\boldsymbol{A}_3$, as well as the rotation $\boldsymbol{R}$, are all functions of position.

Because $\boldsymbol{C}=\boldsymbol{U}^2$, and $\boldsymbol{\gamma}=\frac{1}{2}(\boldsymbol{C}-\boldsymbol{I})$, the principal directions of $\boldsymbol{C}$ and $\boldsymbol{\gamma}$ coincide with those of $\boldsymbol{U}$, and their principal values are λ_i^2 and $\frac{1}{2}(\lambda_i^2-1)$ $(i=1,2,3)$ respectively. Similarly, the principal directions of $\boldsymbol{B}$ and $\boldsymbol{\eta}$ coincide with those of $\boldsymbol{V}$ and their principal values are λ_i^2 and $\frac{1}{2}(1-\lambda_i^{-2})$ $(i=1,2,3)$ respectively. For a given $\boldsymbol{F}$ it is much easier to calculate $\boldsymbol{C}$ or $\boldsymbol{B}$ than $\boldsymbol{U}$ or $\boldsymbol{V}$, and so the easiest way to calculate the principal stretches and principal directions is by calculating the principal values and principal directions of $\boldsymbol{C}$ or $\boldsymbol{B}$.

The principal stretches and principal axes of the deformation tensors can be interpreted in another way. We recall the formula (6.29),

$$\lambda^2=A_RA_SC_{RS} \tag{9.24}$$

For a given tensor $\boldsymbol{C}$ this determines an extension ratio λ for each set of direction cosines A_S in the reference configuration. We enquire for what directions $\boldsymbol{A}$ this extension ratio takes extremal values; thus we seek extremal values of $A_RA_SC_{RS}$, subject to the constraint $A_RA_R=1$. These extremal values are given by the solutions of the equations

$$\frac{\partial}{\partial A_P}\{A_RA_SC_{RS}-\mu^2(A_RA_R-1)\}=0$$

where μ^2 is a Lagrangian multiplier. Since $\partial A_R/\partial A_P=\delta_{RP}$, and $\partial A_S/\partial A_P=\delta_{SP}$, this equation reduces to

$$(C_{PR}-\mu^2\delta_{PR})A_R=0 \tag{9.25}$$

Hence the directions $\boldsymbol{A}$ for which λ^2 is extremal are two of the principal directions of $\boldsymbol{C}$. Therefore the corresponding values of λ^2 are the largest and smallest principal values of $\boldsymbol{C}$, namely λ_1^2 and λ_3^2. A similar procedure applied to the tensor $\boldsymbol{B}$ shows that λ^2 takes its extremal values λ_1^2 and λ_3^2 for directions in the deformed configuration which coincide with two of the principal directions of $\boldsymbol{B}$.

9.4 Strain invariants

It follows from the discussion of Sections 3.8 and 9.3 that the principal stretches λ_1, λ_2 and λ_3 are invariants which are intrinsic to the deformation. Since λ_1, λ_2 and λ_3 are principal values of $\boldsymbol{U}$ and $\boldsymbol{V}$, three symmetric functions of λ_1, λ_2 and λ_3 may be chosen as the basic invariants of $\boldsymbol{U}$ and $\boldsymbol{V}$. However, it is preferable to make use of the fact that λ_1^2, λ_2^2 and λ_3^2 are principal values of $\boldsymbol{C}$ and $\boldsymbol{B}$, and to define the *strain invariants* I_1, I_2 and I_3 as follows:

$$I_1 = \lambda_1^2 + \lambda_2^2 + \lambda_3^2, \qquad I_2 = \lambda_2^2\lambda_3^2 + \lambda_3^2\lambda_1^2 + \lambda_1^2\lambda_2^2, \qquad I_3 = \lambda_1^2\lambda_2^2\lambda_3^2 \tag{9.26}$$

The advantage of this procedure is that $\boldsymbol{C}$ and $\boldsymbol{B}$ are much more easily calculated from $\boldsymbol{F}$ than are $\boldsymbol{U}$ and $\boldsymbol{V}$. The choice (9.26) of the strain invariants is, of course, not unique, but it is one which has proved to be convenient.

Since λ_1^2, λ_2^2 and λ_3^2 are the principal values of both $\boldsymbol{C}$ and $\boldsymbol{B}$, there follow from (3.56) and (3.57):

$$\begin{aligned} I_1 &= \operatorname{tr} \boldsymbol{C} = \operatorname{tr} \boldsymbol{B} = C_{RR} = B_{ii} \\ I_2 &= \tfrac{1}{2}\{(\operatorname{tr} \boldsymbol{C})^2 - \operatorname{tr} \boldsymbol{C}^2\} = \tfrac{1}{2}\{(\operatorname{tr} \boldsymbol{B})^2 - \operatorname{tr} \boldsymbol{B}^2\} \\ &= \tfrac{1}{2}\{C_{RR}C_{SS} - C_{RS}C_{RS}\} = \tfrac{1}{2}\{B_{ii}B_{jj} - B_{ij}B_{ij}\} \\ I_3 &= \det \boldsymbol{C} = \det \boldsymbol{B} \end{aligned} \tag{9.27}$$

Alternative expressions for I_3 are obtained by substituting $\boldsymbol{C}$ and $\boldsymbol{B}$ for $\boldsymbol{A}$ in (3.59).

From (3.58), the Cayley–Hamilton theorem for $\boldsymbol{C}$ and for $\boldsymbol{B}$ can be expressed as

$$\begin{aligned} \boldsymbol{C}^3 - I_1\boldsymbol{C}^2 + I_2\boldsymbol{C} - I_3\boldsymbol{I} &= \boldsymbol{0} \\ \boldsymbol{B}^3 - I_1\boldsymbol{B}^2 + I_2\boldsymbol{B} - I_3\boldsymbol{I} &= \boldsymbol{0} \end{aligned} \tag{9.28}$$

The eigenvalues of $\boldsymbol{C}^{-1}$ and of $\boldsymbol{B}^{-1}$ are λ_1^{-2}, λ_2^{-2} and λ_3^{-2}.

Therefore

$$\begin{aligned}\operatorname{tr} \boldsymbol{C}^{-1} &= \operatorname{tr} \boldsymbol{B}^{-1} = \lambda_1^{-2} + \lambda_2^{-2} + \lambda_3^{-2} \\ &= \lambda_1^{-2}\lambda_2^{-2}\lambda_3^{-2}(\lambda_2^2\lambda_3^2 + \lambda_3^2\lambda_1^2 + \lambda_1^2\lambda_2^2) \\ &= I_2 I_3^{-1}\end{aligned}$$

Hence we obtain the alternative expressions for I_2:

$$I_2 = I_3 \operatorname{tr} \boldsymbol{C}^{-1} = I_3 \operatorname{tr} \boldsymbol{B}^{-1} \tag{9.29}$$

We note also that, from (7.8),

$$I_3 = \det \boldsymbol{C} = \det \boldsymbol{F}^{\mathrm{T}}\boldsymbol{F} = (\det \boldsymbol{F})^2 = \left(\frac{\mathrm{d}v}{\mathrm{d}V}\right)^2 = \left(\frac{\rho_0}{\rho}\right)^2 \tag{9.30}$$

If the material is incompressible then (Section 7.2) $\det \boldsymbol{F} = 1$, and so $I_3 = 1$. Hence in any deformation of an incompressible material, $\lambda_1\lambda_2\lambda_3 = 1$.

Example 9.1 Uniform extensions. For the uniform extensions defined by (6.42) the polar decomposition is trivial: we have $\boldsymbol{F} = \boldsymbol{U} = \boldsymbol{V}$, $\boldsymbol{R} = \boldsymbol{I}$. The principal stretches are λ_1, λ_2 and λ_3 and the coordinate axes are the principal axes of both $\boldsymbol{C}$ and $\boldsymbol{B}$. The strain invariants are

$$I_1 = \lambda_1^2 + \lambda_2^2 + \lambda_3^2, \qquad I_2 = \lambda_2^2\lambda_3^2 + \lambda_3^2\lambda_1^2 + \lambda_1^2\lambda_2^2, \qquad I_3 = \lambda_1^2\lambda_2^2\lambda_3^2$$

Example 9.2 Simple shear. A simple shearing motion is defined by (6.44). From (6.45) and (9.27) the strain invariants for this motion are

$$I_1 = 3 + \tan^2\gamma, \qquad I_2 = 3 + \tan^2\gamma, \qquad I_3 = 1$$

Since $I_3 = 1$, a simple shearing motion is possible in an incompressible material, as is obvious from Fig. 6.4. By calculating the eigenvalues of the matrix of the components of the tensor $\boldsymbol{C}$ given in (6.45), we find that

$$\lambda_1 = \sec\beta + \tan\beta, \qquad \lambda_2 = 1, \qquad \lambda_3 = \sec\beta - \tan\beta$$

where $\tan\beta = \frac{1}{2}\tan\gamma$. The principal directions of $\boldsymbol{C}$ are given by the eigenvectors of the matrix of the components of $\boldsymbol{C}$; these eigenvectors have the following components:

$$\left\{\frac{1}{\sqrt{2}}(1-\sin\beta)^{\frac{1}{2}},\ \frac{1}{\sqrt{2}}(1+\sin\beta)^{\frac{1}{2}},\ 0\right\}^{\mathrm{T}}, \qquad \{0,\ 0,\ 1\}^{\mathrm{T}},$$

$$\left\{\frac{1}{\sqrt{2}}(1+\sin\beta)^{\frac{1}{2}},\ -\frac{1}{\sqrt{2}}(1-\sin\beta)^{\frac{1}{2}},\ 0\right\}^{\mathrm{T}}$$

Similarly, the components of the eigenvectors of $\boldsymbol{B}$ are

$$\left\{\frac{1}{\sqrt{2}}(1+\sin\beta)^{\frac{1}{2}},\ \frac{1}{\sqrt{2}}(1-\sin\beta)^{\frac{1}{2}},\ 0\right\}^{\mathrm{T}},\qquad \{0,\ 0,\ 1\}^{\mathrm{T}},$$

$$\left\{\frac{1}{\sqrt{2}}(1-\sin\beta)^{\frac{1}{2}},\ -\frac{1}{\sqrt{2}}(1+\sin\beta)^{\frac{1}{2}},\ 0\right\}^{\mathrm{T}}$$

The components of the tensor $\boldsymbol{R}$ can be calculated by using the property that $\boldsymbol{R}$ represents the rotation which rotates the orthogonal triad of principal axes of $\boldsymbol{C}$ into the orthogonal triad of principal axes of $\boldsymbol{B}$. Thus if

$$\mathbf{M}_1=\frac{1}{\sqrt{2}}\begin{pmatrix}(1-\sin\beta)^{\frac{1}{2}} & 0 & (1+\sin\beta)^{\frac{1}{2}}\\ (1+\sin\beta)^{\frac{1}{2}} & 0 & -(1-\sin\beta)^{\frac{1}{2}}\\ 0 & \sqrt{2} & 0\end{pmatrix}$$

$$\mathbf{M}_2=\frac{1}{\sqrt{2}}\begin{pmatrix}(1+\sin\beta)^{\frac{1}{2}} & 0 & (1-\sin\beta)^{\frac{1}{2}}\\ (1-\sin\beta)^{\frac{1}{2}} & 0 & -(1+\sin\beta)^{\frac{1}{2}}\\ 0 & \sqrt{2} & 0\end{pmatrix}$$

then $\mathbf{M}_2=\mathbf{R}\mathbf{M}_1$, where $\mathbf{R}$ is the matrix of components of $\boldsymbol{R}$. Since $\mathbf{M}_1$ is orthogonal, it follows that $\mathbf{R}=\mathbf{M}_2\mathbf{M}_1^{\mathrm{T}}$, which gives

$$\mathbf{R}=\begin{pmatrix}\cos\beta & \sin\beta & 0\\ -\sin\beta & \cos\beta & 0\\ 0 & 0 & 1\end{pmatrix}$$

Thus $\boldsymbol{R}$ represents a rotation through β about the X_3-axis. The components of the tensor $\boldsymbol{U}$ are then determined by the equation $\boldsymbol{U}=\boldsymbol{R}^{\mathrm{T}}\cdot\boldsymbol{F}$, which gives

$$(U_{RS})=\begin{pmatrix}\cos\beta & \sin\beta & 0\\ \sin\beta & \sec\beta\,(1+\sin^2\beta) & 0\\ 0 & 0 & 1\end{pmatrix}$$

An alternative procedure is to calculate $\boldsymbol{U}$ directly from the relation $\boldsymbol{U}^2=\boldsymbol{C}$ and to use the relation $\boldsymbol{R}=\boldsymbol{F}\cdot\boldsymbol{U}^{-1}$ to determine $\boldsymbol{R}$.

9.5 Alternative stress measures

In Section 5.2 we defined the component T_{ij} of the Cauchy stress tensor $\boldsymbol{T}$ as the component in the x_j direction of the surface traction on a surface element which is normal to the x_i direction in the current configuration. For some purposes it is more convenient to use a stress tensor which is defined in terms of the traction on a material surface which is specified in the reference configuration.

Consider an element of a material surface which in the reference configuration is normal to the X_R-axis and has area δS. The unit normal to the surface is therefore $\boldsymbol{e}_R$ in the reference configuration. After the deformation (6.1), this element has area δs and unit normal $\boldsymbol{n}_R$ where, from (9.9),

$$\boldsymbol{n}_R = (\det \boldsymbol{F})\frac{\mathrm{d}S}{\mathrm{d}s}\boldsymbol{e}_R \cdot \boldsymbol{F}^{-1} \tag{9.31}$$

The force on this deformed surface is denoted by $\boldsymbol{\pi}_R\,\delta S$. The vector $\boldsymbol{\pi}_R$ is resolved into components Π_{Ri}, so that

$$\boldsymbol{\pi}_R = \Pi_{Ri}\boldsymbol{e}_i \tag{9.32}$$

Thus Π_{Ri} represents the component in the x_i direction of the force on a surface which is normal to the X_R-axis in the reference configuration, measured per unit surface area in the reference configuration.

To relate Π_{Ri} to T_{ij}, we note that the force on the deformed surface element is also equal to $\boldsymbol{n}_R \cdot \boldsymbol{T}\,\delta s$. Hence, from (9.31) and (9.32),

$$\Pi_{Ri}\boldsymbol{e}_i\,\delta S = (\det \boldsymbol{F})\frac{\mathrm{d}S}{\mathrm{d}s}\boldsymbol{e}_R \cdot \boldsymbol{F}^{-1} \cdot \boldsymbol{T}\,\delta s \tag{9.33}$$

Therefore, by equating components on either side of (9.33) and taking the limit as $\delta S \to 0$, we obtain

$$\Pi_{Ri} = (\det \boldsymbol{F})F^{-1}_{Rj}T_{ij} \tag{9.34}$$

Hence Π_{Ri} are components of a second-order tensor $\boldsymbol{\Pi}$, where

$$\boldsymbol{\Pi} = (\det \boldsymbol{F})\boldsymbol{F}^{-1} \cdot \boldsymbol{T} \tag{9.35}$$

and conversely

$$\boldsymbol{T} = (\det \boldsymbol{F})^{-1}\boldsymbol{F} \cdot \boldsymbol{\Pi} \tag{9.36}$$

The tensor $\boldsymbol{\Pi}$ is not symmetric. We shall call it the *nominal stress tensor*. It is often also called the *first Piola–Kirchhoff* stress tensor, but some authors reserve this term for its transpose, $\boldsymbol{\Pi}^{\mathrm{T}}$.

By considering the equilibrium of an elementary tetrahedron, three of whose faces are normal to the coordinate axes in the reference configuration, it can be shown that the traction $\boldsymbol{t}^{(N)}$ (measured per unit area in the reference configuration) on a material surface which has unit normal $\boldsymbol{N}$ in the reference configuration is given by

$$\boldsymbol{t}^{(N)} = \boldsymbol{N} \cdot \boldsymbol{\Pi} \tag{9.37}$$

By considering the resultant surface and body forces on an arbitrary region of a body, and referred to the body in its reference configuration, the equations of motion can be expressed in the form

$$\frac{\partial \Pi_{Ri}}{\partial X_R} + \rho_0 b_i = \rho_0 f_i \tag{9.38}$$

The *second Piola–Kirchhoff stress tensor* $\boldsymbol{P}$ is defined as

$$\boldsymbol{P} = \boldsymbol{\Pi} \cdot (\boldsymbol{F}^{-1})^{\mathrm{T}} = (\det \boldsymbol{F})\boldsymbol{F}^{-1} \cdot \boldsymbol{T} \cdot (\boldsymbol{F}^{-1})^{\mathrm{T}} \tag{9.39}$$

Hence

$$\boldsymbol{\Pi} = \boldsymbol{P} \cdot \boldsymbol{F}^{\mathrm{T}}, \qquad \boldsymbol{T} = (\det \boldsymbol{F})^{-1} \boldsymbol{F} \cdot \boldsymbol{P} \cdot \boldsymbol{F}^{\mathrm{T}} \tag{9.40}$$

The tensor $\boldsymbol{P}$ is symmetric. It does not have any simple direct interpretation.

The traction on a surface defined in the current configuration is not determined by $\boldsymbol{\Pi}$ or $\boldsymbol{P}$ unless $\boldsymbol{F}$ is also given. To leading order, $\boldsymbol{\Pi}$ and $\boldsymbol{P}$ reduce to $\boldsymbol{T}$ in the case of infinitesimal displacement gradients. We shall not use $\boldsymbol{\Pi}$ or $\boldsymbol{P}$ in this book, except to point out in Section 10.2 that certain constitutive equations can be expressed concisely in terms of $\boldsymbol{\Pi}$ and $\boldsymbol{P}$.

9.6 Problems

1. For the deformation defined in Chapter 6, Problem 2, find: (a) the direction of the normal to a material surface element in the deformed configuration which had normal direction (1, 1, 1) in the reference configuration; (b) the ratio of the areas of this surface element in the reference and deformed configurations; (c) the principal stretches; (d) the principal axes of $\boldsymbol{C}$ and of $\boldsymbol{B}$.

2. Determine C_{RS} for the deformation given by

$$x_1 = \frac{a(X_1^2 - X_2^2)}{(X_1^2 + X_2^2)^{\frac{1}{2}}}, \qquad x_2 = \frac{2aX_1X_2}{(X_1^2 + X_2^2)^{\frac{1}{2}}}, \qquad x_3 = bX_3$$

where a and b are constants. Find the principal stretches and the principal axes of $\boldsymbol{C}$.

3. For the deformation defined by

$$X_1=\tfrac{1}{2}A(x_1^2+x_2^2) \qquad X_2=\lambda A^{-1}\tan^{-1}(x_2/x_1), \qquad X_3=\lambda^{-1}x_3$$

where A and λ are constants, find B_{ij}^{-1}. Prove that the squares of the principal stretches are λ^2 and the two roots of the quadratic equation $\mu^2\lambda^2-\mu(A^2r^2+\lambda^2A^{-2}r^{-2})+1=0$ where $r^2=x_1^2+x_2^2$. Hence show that $\det \mathbf{B}^{-1}=1$.

4. For the homogeneous deformation

$$x_1=\alpha X_1+\beta X_2, \qquad x_2=-\alpha X_1+\beta X_2, \qquad x_3=\mu X_3$$

where α, β and μ are positive constants, determine the components C_{RS} and the principal stretches, and find $\boldsymbol{R}$ and $\boldsymbol{U}$ for the polar decomposition $\boldsymbol{F}=\boldsymbol{R}\cdot\boldsymbol{U}$.

5. A fluid moves so that the particle at the point with coordinates (X_1, X_2, X_3) at time $t=0$ is at the point with coordinates $(x_1(\tau), x_2(\tau), x_3(\tau))$ at time $t=\tau$, where

$$x_1(\tau)=X_1+\alpha\tau X_2+\alpha\beta\tau^2X_3, \qquad x_2(\tau)=X_2+2\beta\tau X_3, \qquad x_3(\tau)=X_3$$

and α and β are constants. Obtain expressions for $x_i(\tau)$ in terms of the coordinates x_i of the particle at time t and determine the components of the tensor $\boldsymbol{C}(\tau)$ defined by

$$C_{ij}(\tau)=\frac{\partial x_k(\tau)}{\partial x_i}\frac{\partial x_k(\tau)}{\partial x_j}$$

By expanding $\boldsymbol{C}(\tau)$ as a power series in $s=t-\tau$, obtain the Rivlin–Ericksen tensors $\boldsymbol{A}^{(n)}(t)$ for all values of n, where

$$\boldsymbol{A}^{(n)}(t)=(\mathrm{d}^n\boldsymbol{C}(\tau)/\mathrm{d}\tau^n)_{\tau=t}$$

6. The Rivlin–Ericksen tensors $\boldsymbol{A}^{(n)}$ satisfy the relations

$$A_{ij}^{(0)}=\delta_{ij}, \qquad A_{ij}^{(1)}=2D_{ij}, \qquad A_{ij}^{(n+1)}=\frac{\mathrm{D}}{\mathrm{D}t}A_{ij}^{(n)}+A_{ik}^{(n)}\frac{\partial v_k}{\partial x_j}+A_{kj}^{(n)}\frac{\partial v_k}{\partial x_i}$$

Evaluate these tensors for the steady flow $v_1=v(x_2)$, $v_2=0$, $v_3=0$, showing that $A_{ij}^{(n)}=0$ for $n\geq 3$.

10

Non-linear constitutive equations

10.1 Nonlinear theories

In Chapter 8 we discussed some of the linear theories of continuum mechanics. Linearity of the governing equations is always a great advantage in the solution of boundary-value problems, because it enables the techniques of linear analysis to be employed. As a result of this, the linear theories of continuum mechanics have been highly developed and applied to numerous problems. Many common materials are adequately modelled by linear constitutive equations. However, there are also many materials whose mechanical behaviour is strongly non-linear, and to describe this behaviour it is essential to formulate appropriate non-linear constitutive equations. We give some examples in this chapter.

10.2 The theory of finite elastic deformations

The linear theory of elasticity which was formulated in Section 8.3 is very effective for many purposes. However, because it is restricted to the case in which the deformation gradients are small, it has limitations. For example, the linear theory is inadequate to describe the mechanical behaviour of materials such as rubber, which are capable of undergoing large deformations, but (to a good approximation) behave elastically in the sense described in Section 8.3. To model the behaviour of rubber-like materials, and for other purposes, we require a theory of finite elastic deformations.

To formulate a theory of finite elastic deformations we postulate, as in Section 8.3, the existence of a strain-energy function $W=\rho_0 e$ which depends only on the deformation and has the property (b) (p. 111). Thus equation (8.12) remains valid in the finite theory of elasticity. However, it is no longer assumed that

W may be approximated by a quadratic function of the infinitesimal strain components. Instead, we permit W to depend in an arbitrary manner on the deformation gradient components F_{iR}, so that (8.7) is replaced by the more general relation

$$W = W(F_{iR}) = W(\boldsymbol{F}) \tag{10.1}$$

Then (6.76), (8.12) and (10.1) give

$$T_{ij}\frac{\partial v_i}{\partial x_j} = \frac{\rho}{\rho_0}\frac{\mathrm{D}W}{\mathrm{D}t} = \frac{\rho}{\rho_0}\frac{\partial W}{\partial F_{iR}}\frac{\mathrm{D}F_{iR}}{\mathrm{D}t} = \frac{\rho}{\rho_0}\frac{\partial W}{\partial F_{iR}}\frac{\partial x_j}{\partial X_R}\frac{\partial v_i}{\partial x_j}$$

This relation is valid for all values of $\partial v_i/\partial x_j$, and so

$$T_{ij} = \frac{\rho}{\rho_0}\frac{\partial x_j}{\partial X_R}\frac{\partial W}{\partial F_{iR}} = \frac{\rho}{\rho_0}F_{jR}\frac{\partial W}{\partial F_{iR}} \tag{10.2}$$

Equation (10.2) is a form of the constitutive equation for finite elasticity. Its apparent simplicity is deceptive, because it requires W to be expressed as a function of the nine components F_{iR}. It would clearly be impracticable to perform experiments to determine this function for any particular elastic material.

The value of the strain–energy function is not changed if a rigid-body rotation is superposed on the deformation. Suppose that a typical particle initially has position vector $\boldsymbol{X}$, and that in a motion it moves to the point with position vector $\boldsymbol{x}$. In a further superposed rigid-body rotation the particle originally at $\boldsymbol{X}$ moves to $\bar{\boldsymbol{x}} = \boldsymbol{M}\cdot\boldsymbol{x}$, where $\boldsymbol{M}$ is a proper orthogonal tensor. Let

$$F_{iR} = \frac{\partial x_i}{\partial X_R}, \qquad \bar{F}_{iR} = \frac{\partial \bar{x}_i}{\partial X_R}$$

Then

$$\bar{F}_{iR} = M_{ij}\frac{\partial x_j}{\partial X_R} = M_{ij}F_{jR}, \qquad \text{or} \qquad \bar{\boldsymbol{F}} = \boldsymbol{M}\cdot\boldsymbol{F} \tag{10.3}$$

Then we require that

$$W(\boldsymbol{F}) = W(\boldsymbol{M}\cdot\boldsymbol{F}) \tag{10.4}$$

for all proper orthogonal tensors $\boldsymbol{M}$. Equation (10.4) is a restriction on the manner in which W may depend on $\boldsymbol{F}$. To make this restriction explicit we employ the polar decomposition theorem to express (10.4) in the form

$$W(\boldsymbol{F}) = W(\boldsymbol{M}\cdot\boldsymbol{R}\cdot\boldsymbol{U})$$

Since this relation holds for all proper orthogonal tensors $\boldsymbol{M}$, it holds in particular when $\boldsymbol{M}=\boldsymbol{R}^{\mathrm{T}}$. Hence

$$W(\boldsymbol{F})=W(\boldsymbol{U})$$

Thus W can be expressed as a function of the six components of the symmetric tensor $\boldsymbol{U}$. However, there is a one-to-one correspondence between the tensors $\boldsymbol{U}$ and $\boldsymbol{C}$ (Section 9.2) and so equivalently (and more conveniently) we may regard W as a function of the six components C_{RS} of $\boldsymbol{C}$. Consequently a necessary condition for W to be independent of superposed rigid-body motions is that W can be expressed in the form

$$W=W(\boldsymbol{C}) \tag{10.5}$$

where of course, the function W is not the same in (10.5) as it is in (10.1). Because $\boldsymbol{C}$ does not change its value in a superposed rigid-body motion, the form (10.5) is also sufficient to ensure that W remains unchanged in a superimposed rigid-body motion, and so no further simplifications can be achieved in this way.

When W is expressed in the form (10.5), we have

$$\begin{aligned}\frac{\mathrm{D}W}{\mathrm{D}t}&=\frac{\partial W}{\partial C_{RS}}\frac{\mathrm{D}C_{RS}}{\mathrm{D}t}=\frac{\partial W}{\partial C_{RS}}\frac{\mathrm{D}}{\mathrm{D}t}\left(\frac{\partial x_i}{\partial X_R}\frac{\partial x_i}{\partial X_S}\right)\\&=\frac{\partial W}{\partial C_{RS}}\left(\frac{\partial v_i}{\partial X_R}\frac{\partial x_i}{\partial X_S}+\frac{\partial x_i}{\partial X_R}\frac{\partial v_i}{\partial X_S}\right)\end{aligned}$$

By interchanging the dummy indices R and S in one of the terms on the right-hand side, this gives

$$\frac{\mathrm{D}W}{\mathrm{D}t}=\left(\frac{\partial W}{\partial C_{RS}}+\frac{\partial W}{\partial C_{SR}}\right)\frac{\partial x_i}{\partial X_R}\frac{\partial v_i}{\partial X_S}=\left(\frac{\partial W}{\partial C_{RS}}+\frac{\partial W}{\partial C_{SR}}\right)\frac{\partial x_i}{\partial X_R}\frac{\partial x_j}{\partial X_S}\frac{\partial v_i}{\partial x_j} \tag{10.6}$$

In (10.6), and subsequently, W is regarded as a symmetric function of C_{RS} and C_{SR}, although these components are equal to each other. Since $\partial v_i/\partial x_j$ is arbitrary, (8.12) and (10.6) now give

$$T_{ij}=\frac{\rho}{\rho_0}\frac{\partial x_i}{\partial X_R}\frac{\partial x_j}{\partial X_S}\left(\frac{\partial W}{\partial C_{RS}}+\frac{\partial W}{\partial C_{SR}}\right) \tag{10.7}$$

This is the required general form of the constitutive equation for a finite elastic solid.

We note in passing that the constitutive equations (10.2) and (10.7) take simpler forms when they are expressed in terms of the nominal or Piola–Kirchhoff stress tensors. Since $\rho_0/\rho = \det \boldsymbol{F}$, we have from (9.35) and (10.2),

$$\Pi_{Ri} = \partial W / \partial F_{iR}$$

and from (9.39) and (10.7)

$$P_{RS} = \frac{\partial W}{\partial C_{RS}} + \frac{\partial W}{\partial C_{SR}}$$

Any material symmetries which the material possesses will restrict the manner in which W may depend upon $\boldsymbol{C}$. Suppose, for example, that the proper orthogonal matrix $\boldsymbol{Q}$ defines a rotational symmetry of the material. The effect of replacing the deformation (8.1) by the deformation (8.2) is to replace $\boldsymbol{F}$ by $\boldsymbol{Q}^{\mathrm{T}} \cdot \boldsymbol{F} \cdot \boldsymbol{Q}$, and so to replace $\boldsymbol{C} = \boldsymbol{F}^{\mathrm{T}} \cdot \boldsymbol{F}$ by $\boldsymbol{Q}^{\mathrm{T}} \cdot \boldsymbol{C} \cdot \boldsymbol{Q}$. However, when $\boldsymbol{Q}$ defines a rotational symmetry, this replacement leaves the value of W unchanged. Thus

$$W(\boldsymbol{C}) = W(\boldsymbol{Q}^{\mathrm{T}} \cdot \boldsymbol{C} \cdot \boldsymbol{Q}) \tag{10.8}$$

for all rotational symmetries $\boldsymbol{Q}$. Similarly, if $\boldsymbol{R}$ defines a reflectional symnetry, then

$$W(\boldsymbol{C}) = W(\boldsymbol{R}^{\mathrm{T}} \cdot \boldsymbol{C} \cdot \boldsymbol{R}) \tag{10.9}$$

If the material is *isotropic*, then (10.8) holds for *all* rotations $\boldsymbol{Q}$. Then (10.8) can be interpreted as a statement that W, regarded as a function of C_{RS}, takes the same form in any coordinate system, so that (Section 3.8) W is an invariant of $\boldsymbol{C}$. Three independent invariants of $\boldsymbol{C}$ are the strain invariants I_1, I_2 and I_3 defined by (9.26) or (9.27); it can be shown that any invariant of $\boldsymbol{C}$ can be expressed as a function of I_1, I_2 and I_3. Hence, for an isotropic material, W can be expressed in the form

$$W = W(I_1, I_2, I_3) \tag{10.10}$$

where again the function W is a different function from that in (10.1) and (10.5). It can be verified that if W has the form (10.10), it also satisfies the condition (10.9) for all reflections $\boldsymbol{R}$.

When W has the form (10.10), we have

$$\frac{\partial W}{\partial C_{RS}} = \frac{\partial W}{\partial I_1}\frac{\partial I_1}{\partial C_{RS}} + \frac{\partial W}{\partial I_2}\frac{\partial I_2}{\partial C_{RS}} + \frac{\partial W}{\partial I_3}\frac{\partial I_3}{\partial C_{RS}} \tag{10.11}$$

From (9.27), it follows that

$$\frac{\partial I_1}{\partial C_{RS}} = \frac{\partial C_{PP}}{\partial C_{RS}} = \delta_{PR}\delta_{PS} = \delta_{RS}$$

$$\begin{aligned}\frac{\partial I_2}{\partial C_{RS}} &= \frac{1}{2}\frac{\partial}{\partial C_{RS}}\{C_{PP}C_{QQ} - C_{PQ}C_{PQ}\} \\ &= \tfrac{1}{2}\{\delta_{PR}\delta_{PS}C_{QQ} + C_{PP}\delta_{RQ}\delta_{SQ} - 2C_{PQ}\delta_{PR}\delta_{QS}\} \\ &= I_1\delta_{RS} - C_{RS}\end{aligned} \tag{10.12}$$

The expression for $\partial I_3/\partial C_{RS}$ is most easily obtained by taking the trace of (9.28), which gives

$$I_3 = \tfrac{1}{3}\{\operatorname{tr}\boldsymbol{C}^3 - I_1 \operatorname{tr}\boldsymbol{C}^2 + I_2 \operatorname{tr}\boldsymbol{C}\} \tag{10.13}$$

and from this it follows that

$$\frac{\partial I_3}{\partial C_{RS}} = I_2\delta_{RS} - I_1 C_{RS} + C_{RP}C_{SP} \tag{10.14}$$

By substituting from (10.11), (10.12) and (10.14) into (10.7), we obtain

$$T_{ij} = 2\frac{\rho}{\rho_0}\frac{\partial x_i}{\partial X_R}\frac{\partial x_j}{\partial X_S} \times\left\{\left(\frac{\partial W}{\partial I_1} + I_1\frac{\partial W}{\partial I_2} + I_2\frac{\partial W}{\partial I_3}\right)\delta_{RS} - \left(\frac{\partial W}{\partial I_2} + I_1\frac{\partial W}{\partial I_3}\right)C_{RS} + \frac{\partial W}{\partial I_3}C_{RP}C_{SP}\right\}$$

This is a form of the constitutive equation for an *isotropic* finite elastic solid. It may be expressed more concisely using tensor notation as

$$\boldsymbol{T} = 2(I_3)^{-\frac{1}{2}}\boldsymbol{F}\cdot\{(W_1 + I_1W_2 + I_2W_3)\boldsymbol{I} - (W_2 + I_1W_3)\boldsymbol{C} + W_3\boldsymbol{C}^2\}\cdot\boldsymbol{F}^{\mathrm{T}} \tag{10.15}$$

where we have used the relation $I_3 = (\rho_0/\rho)^2$, and for brevity we have introduced the notations

$$W_1 = \partial W/\partial I_1, \qquad W_2 = \partial W/\partial I_2, \qquad W_3 = \partial W/\partial I_3 \tag{10.16}$$

Equation (10.15) may be further simplified by noting, from (6.27) and (6.33) that

$$\boldsymbol{F}\cdot\boldsymbol{F}^{\mathrm{T}} = \boldsymbol{B}, \qquad \boldsymbol{F}\cdot\boldsymbol{C}\cdot\boldsymbol{F}^{\mathrm{T}} = \boldsymbol{B}^2, \qquad \boldsymbol{F}\cdot\boldsymbol{C}^2\cdot\boldsymbol{F}^{\mathrm{T}} = \boldsymbol{B}^3$$

and hence that (10.15) may be written as

$$\boldsymbol{T} = 2(I_3)^{-\frac{1}{2}}\{(W_1 + I_1W_2 + I_2W_3)\boldsymbol{B} - (W_2 + I_1W_3)\boldsymbol{B}^2 + W_3\boldsymbol{B}^3\}$$

We now use (9.28) to eliminate $\boldsymbol{B}^3$. This gives

$$\boldsymbol{T}=2(I_3)^{-\frac{1}{2}}\{I_3 W_3 \boldsymbol{I}+(W_1+I_1 W_2)\boldsymbol{B}-W_2\boldsymbol{B}^2\} \tag{10.17}$$

Also, by multiplying the second equation of (9.28) by $\boldsymbol{B}^{-1}$, we have

$$\boldsymbol{B}^2-I_1\boldsymbol{B}+I_2\boldsymbol{I}-I_3\boldsymbol{B}^{-1}=\boldsymbol{0}$$

and so $\boldsymbol{B}^2$ can be eliminated from (10.17) in favour of $\boldsymbol{B}^{-1}$, which gives

$$\boldsymbol{T}=2(I_3)^{-\frac{1}{2}}\{(I_2 W_2+I_3 W_3)\boldsymbol{I}+W_1\boldsymbol{B}-I_3 W_2\boldsymbol{B}^{-1}\} \tag{10.18}$$

In practice, (10.17) and (10.18) are the forms of the constitutive equation for an isotropic elastic solid which are found to be most convenient.

Further simplification arises if the material is *incompressible.* In this case $I_3=1$, but it is not sufficient to set $I_3=1$ in the constitutive equation, because in the limiting case of an incompressible material certain derivatives of W tend to infinity. The difficulty is most easily avoided by introduced an arbitrary Lagrangian multiplier $-\frac{1}{2}p$ and writing W in the form

$$W=W(I_1, I_2)-\tfrac{1}{2}p(I_3-1) \tag{10.19}$$

The analysis leading to (10.17) and (10.18) then goes through as before, but I_3 takes the value one and W_3 is replaced by $-\frac{1}{2}p$. Since p is undetermined, the other terms multiplying $\boldsymbol{I}$ in (10.17) and (10.18) may be absorbed into p, so that for an incompressible isotropic finite elastic solid the constitutive equation can be expressed in either of the forms

$$\begin{aligned}\boldsymbol{T}&=-p\boldsymbol{I}+2(W_1+I_1 W_2)\boldsymbol{B}-2W_2\boldsymbol{B}^2\\ \boldsymbol{T}&=-p\boldsymbol{I}+2W_1\boldsymbol{B}-2W_2\boldsymbol{B}^{-1}\end{aligned} \tag{10.20}$$

Incompressibility is an example of a *kinematic constraint.* The mechanical effect of such a constraint is to give rise to a *reaction stress* which does no work in any motion which is compatible with the constraint. In the case of incompressibility, the reaction stress is an arbitrary hydrostatic pressure $-p\boldsymbol{I}$, which is not given by a constitutive equation but can only be determined by using equations of motion (or equilibrium) and boundary conditions. Such an arbitrary hydrostatic pressure must always be included as part of the stress in a body of any incompressible material.

The equations of linear elasticity theory can be recovered from (10.7) by expanding all quantities in powers of the displacement gradients and discarding terms on the right-hand side of (10.7) which are of degree higher than the first in these gradients.

10.3 A non-linear viscous fluid

In Section 8.4 we considered fluids with constitutive equations of the form (8.23), in which $\boldsymbol{T}+p\boldsymbol{I}$ is linear in the rate-of-strain components. This theory proves to be very satisfactory for describing the behaviour of many fluids, including the commonest fluids, air and water, over a very wide range of rates of strain. However, there are also fluids, including blood and many fluids which are important in chemical engineering processes, which exhibit phenomena (which in some cases are quite spectacular), which cannot be explained on the basis of the linear model. Such fluids are described as *non-Newtonian fluids*. For non-Newtonian fluids the assumption that the stress depends linearly on rate of strain is inadequate. Therefore in this section we discard linearity and begin with the assumption that $\boldsymbol{T}$ depends in a general way on density, temperature, and the velocity-gradient tensor. Thus

$$T_{ij} = T_{ij}(\partial v_p/\partial x_q, \rho, \theta) \tag{10.21}$$

or, in tensor notation

$$\boldsymbol{T} = \boldsymbol{T}(\boldsymbol{L}, \rho, \theta) \tag{10.22}$$

We first consider whether the requirements that $\boldsymbol{T}$ is independent of superposed rigid-body motions places any restrictions on (10.22). Since, by (6.72), $\boldsymbol{L} = \boldsymbol{D} + \boldsymbol{W}$, we can replace (10.22) by

$$\boldsymbol{T} = \boldsymbol{T}(\boldsymbol{D}, \boldsymbol{W}, \rho, \theta) \tag{10.23}$$

where $\boldsymbol{T}$ represents a different function on the right-hand side of (10.23) from the function which it represents on the right-hand side of (10.22).

Suppose a body undergoes the motion

$$\boldsymbol{x} = \boldsymbol{x}(\boldsymbol{X}, t), \qquad \boldsymbol{v} = \boldsymbol{v}(\boldsymbol{x}, t) \tag{10.24}$$

Consider a new motion which differs from (10.24) only by a superposed time-dependent rigid rotation, so that at time t the position $\bar{\boldsymbol{x}}$ of the particle initially at $\boldsymbol{X}$ is given by

$$\bar{\boldsymbol{x}} = \boldsymbol{M}(t) \cdot \boldsymbol{x}(\boldsymbol{X}, t) \tag{10.25}$$

where $\boldsymbol{M}$ is a time-dependent proper orthogonal tensor. Then in the second motion the velocity is

$$\bar{\boldsymbol{v}}=\frac{\mathrm{D}\bar{\boldsymbol{x}}}{\mathrm{D}t}=\dot{\boldsymbol{M}}\cdot\boldsymbol{x}+\boldsymbol{M}\cdot\dot{\boldsymbol{x}}=\dot{\boldsymbol{M}}\cdot\boldsymbol{x}+\boldsymbol{M}\cdot\boldsymbol{v} \tag{10.26}$$

The velocity-gradient components in the second motion are given by

$$\bar{L}_{ij}=\frac{\partial\bar{v}_i}{\partial\bar{x}_j}=\frac{\partial\bar{v}_i}{\partial x_k}\frac{\partial x_k}{\partial\bar{x}_j}=\left(\dot{M}_{ip}\delta_{pk}+M_{ip}\frac{\partial v_p}{\partial x_k}\right)M_{jk}$$

or, in tensor notation, as

$$\bar{\boldsymbol{L}}=(\dot{\boldsymbol{M}}+\boldsymbol{M}\cdot\boldsymbol{L})\cdot\boldsymbol{M}^{\mathrm{T}}$$

It follows that the rate-of-strain tensor $\bar{\boldsymbol{D}}$ and the spin tensor $\bar{\boldsymbol{W}}$ for the second motion are given by

$$\left.\begin{aligned}\bar{\boldsymbol{D}}&=\tfrac{1}{2}(\bar{\boldsymbol{L}}+\bar{\boldsymbol{L}}^{\mathrm{T}})=\tfrac{1}{2}(\dot{\boldsymbol{M}}\cdot\boldsymbol{M}^{\mathrm{T}}+\boldsymbol{M}\cdot\dot{\boldsymbol{M}}^{\mathrm{T}})+\tfrac{1}{2}\boldsymbol{M}\cdot(\boldsymbol{L}+\boldsymbol{L}^{\mathrm{T}})\cdot\boldsymbol{M}^{\mathrm{T}}\\ \bar{\boldsymbol{W}}&=\tfrac{1}{2}(\bar{\boldsymbol{L}}-\bar{\boldsymbol{L}}^{\mathrm{T}})=\tfrac{1}{2}(\dot{\boldsymbol{M}}\cdot\boldsymbol{M}^{\mathrm{T}}-\boldsymbol{M}\cdot\dot{\boldsymbol{M}}^{\mathrm{T}})+\tfrac{1}{2}\boldsymbol{M}\cdot(\boldsymbol{L}-\boldsymbol{L}^{\mathrm{T}})\cdot\boldsymbol{M}^{\mathrm{T}}\end{aligned}\right\} \tag{10.27}$$

However, since $\boldsymbol{M}$ is orthogonal, $\boldsymbol{M}\cdot\boldsymbol{M}^{\mathrm{T}}=\boldsymbol{I}$, and it follows that $\dot{\boldsymbol{M}}\cdot\boldsymbol{M}^{\mathrm{T}}+\boldsymbol{M}\cdot\dot{\boldsymbol{M}}^{\mathrm{T}}=\boldsymbol{0}$. Hence (10.27) may be written as

$$\bar{\boldsymbol{D}}=\boldsymbol{M}\cdot\boldsymbol{D}\cdot\boldsymbol{M}^{\mathrm{T}},\qquad \bar{\boldsymbol{W}}=\boldsymbol{M}\cdot(\boldsymbol{M}^{\mathrm{T}}\cdot\dot{\boldsymbol{M}}+\boldsymbol{W})\cdot\boldsymbol{M}^{\mathrm{T}} \tag{10.28}$$

If $\boldsymbol{T}$ is the stress which arises from the first motion, then independence of superposed rotations requires that the second motion gives rise to the stress $\bar{\boldsymbol{T}}=\boldsymbol{M}\cdot\boldsymbol{T}\cdot\boldsymbol{M}^{\mathrm{T}}$. However, from (10.23)

$$\bar{\boldsymbol{T}}=\boldsymbol{T}(\bar{\boldsymbol{D}},\bar{\boldsymbol{W}},\rho,\theta) \tag{10.29}$$

Hence from (10.23), (10.28) and (10.29),

$$\begin{aligned}\boldsymbol{T}\{\boldsymbol{M}\cdot\boldsymbol{D}\cdot\boldsymbol{M}^{\mathrm{T}},\boldsymbol{M}\cdot(\boldsymbol{M}^{\mathrm{T}}\cdot\dot{\boldsymbol{M}}+\boldsymbol{W})\cdot\boldsymbol{M}^{\mathrm{T}},\rho,\theta\}&\\ =\boldsymbol{M}\cdot\boldsymbol{T}\{\boldsymbol{D},\boldsymbol{W},\rho,\theta\}\cdot\boldsymbol{M}^{\mathrm{T}}&\end{aligned} \tag{10.30}$$

and the function $\boldsymbol{T}$ must satisfy this condition identically for all proper orthogonal tensors $\boldsymbol{M}$.

To make (10.30) explicit, we suppose first that $\boldsymbol{M}=\boldsymbol{I}$, $\dot{\boldsymbol{M}}\neq\boldsymbol{0}$. Then (10.30) becomes

$$\boldsymbol{T}\{\boldsymbol{D},\dot{\boldsymbol{M}}+\boldsymbol{W},\rho,\theta\}=\boldsymbol{T}\{\boldsymbol{D},\boldsymbol{W},\rho,\theta\}$$

Hence the value of $\boldsymbol{T}$ is independent of the value of $\boldsymbol{W}$. Therefore the arguments $\boldsymbol{W}$ and $\bar{\boldsymbol{W}}$ may be omitted in (10.23) and

(10.29). Dependence of the stress on the nine components of $\boldsymbol{L}$ can be replaced by dependence on the six components of $\boldsymbol{D}$ (this result was implicitly assumed in Section 8.4). When the argument $\boldsymbol{W}$ is omitted, (10.30) reduces to the form

$$\boldsymbol{T}(\boldsymbol{M}\cdot\boldsymbol{D}\cdot\boldsymbol{M}^{\mathrm{T}}, \rho, \theta) = \boldsymbol{M}\cdot\boldsymbol{T}(\boldsymbol{D}, \rho, \theta)\cdot\boldsymbol{M}^{\mathrm{T}} \tag{10.31}$$

for all orthogonal tensors $\boldsymbol{M}$. A tensor function $\boldsymbol{T}$ with the property (10.31) is said to be an *isotropic tensor function* of $\boldsymbol{D}$. If $\boldsymbol{T}$ is a linear function of $\boldsymbol{D}$, as in Section 8.4, then (10.31) implies that the stress is of the form (8.25), so that the fluid is *necessarily* isotropic. This justifies the statement made in Section 8.4 that it is not essential to introduce isotropy as a separate assumption. The same is true in the general case, for (10.31) can be interpreted as a statement that the material is isotropic.

It is shown in the Appendix that the most general tensor function $\boldsymbol{T}$ which satisfies (10.31) is of the form

$$\boldsymbol{T} = -p\boldsymbol{I} + \alpha\boldsymbol{D} + \beta\boldsymbol{D}^2 \tag{10.32}$$

where p, α and β are functions of ρ, θ and invariants of $\boldsymbol{D}$, namely

$$\mathrm{tr}\,\boldsymbol{D}, \qquad \tfrac{1}{2}\{(\mathrm{tr}\,\boldsymbol{D})^2 - \mathrm{tr}\,\boldsymbol{D}^2\}, \qquad \det \boldsymbol{D}$$

A material with the constitutive equation (10.32) is called a *Reiner–Rivlin fluid.* If the fluid is incompressible then ρ is constant and $\mathrm{tr}\,\boldsymbol{D} = 0$, so that α and β depend only on θ and the second two invariants of $\boldsymbol{D}$, and p represents an arbitrary pressure.

Although the result (10.32) is of mathematical interest, in practice it has been found that markedly non-Newtonian fluids have a more complex behaviour than is permitted by the model defined by (10.21). We discuss a more general class of materials briefly in the next section.

10.4 Non-linear viscoelasticity

In Section 8.5 we outlined the linear theory of viscoelasticity. In a viscoelastic material (which may be a solid or a fluid) the stress depends not only on the current deformation but also on the past history of deformation. The material may be said to have a 'memory'. Linear viscoelasticity is governed by the superposition principle, according to which the effects of past deformations may be superposed to give the present stress. Many non-Newtonian fluids, and many solids (especially polymers) are viscoelastic in that the stress depends on the deformation history, but this dependence is more complicated than a direct superposition of the form (8.29). The modelling of such materials requires the non-linear theory of viscoelasticity.

In a non-linear viscoelastic material the stress at a particle depends not only on the current deformation, but also on the history of the deformation. Thus formally, the constitutive equation may be written as

$$\boldsymbol{T}(t) = \mathscr{T}\left\{ \partial x_i(\tau)/\partial X_R \right\}_{\tau=-\infty}^{\tau=t} \tag{10.33}$$

It can be shown that if $\boldsymbol{T}$ is independent of superposed rigid-body motions, then (10.33) can be reduced to

$$\boldsymbol{T}(t) = \boldsymbol{F} \cdot \mathscr{T}\left\{ C_{RS}(\tau) \right\}_{\tau=-\infty}^{\tau=t} \cdot \boldsymbol{F}^{\mathrm{T}} \tag{10.34}$$

and further reductions can be made if the material has any material symmetry.

In the non-linear case it is no longer possible to use the superposition principle which leads to the comparatively simple integral representation (8.29) for $\boldsymbol{T}$. The functional in (10.34) can be represented, either exactly or approximately, in various ways, but the resulting thoeries are too advanced for consideration here.

10.5 Plasticity

Many materials, particularly metals, conform well to the linear theory of elasticity provided that the stress does not exceed certain limits, but if they are subjected to stress beyond these limits they acquire a permanent deformation which does not disappear when the stress is removed. Since elasticity is a reversible phenomenon, this is clearly inelastic behaviour. It is not a viscoelastic phenomenon, because the viscoelastic stress depends on the rate of deformation, and to a good approximation it is found that although the stress in a metal depends on the previous deformation, it is independent of the rate at which that deformation took place. The phenomenon is called *plasticity*; characteristically

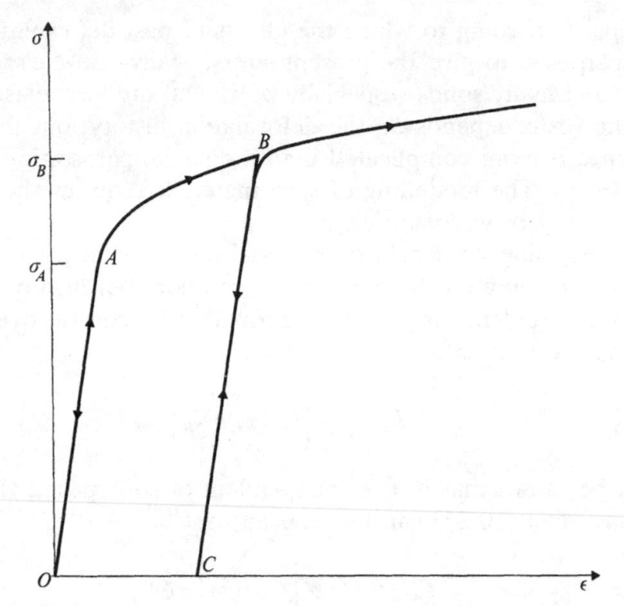

Figure 10.1 Typical stress–strain curve for a plastic solid

it occurs in crystalline materials, and in particular it occurs in the solid metals which are in everyday use, such as steel, aluminium and copper.

Figure 10.1 illustrates the main features of the stress–strain curve in uniaxial tension of a typical metal; the axial stress is denoted by σ and the axial strain by ϵ. For simplicity it is assumed that the strain is sufficiently small for the infinitesimal strain measure to be adequate.

For the deformation which corresponds to the section OA of the curve, the relation between σ and ϵ is, to a good approximation, linear. If the stress is removed before σ reaches the value σ_A, the strain returns to zero. In this range, the behaviour is that of linear elasticity theory. For stress greater than σ_A, the curve departs from a straight line. The stress σ_A is called the *initial yield stress* in tension. The change of slope at A may be abrupt or gradual. If the stress is increased to $\sigma_B > \sigma_A$, and then reduced to zero, the unloading curve BC is followed; to a good approximation, BC is parallel to OA. When the stress is zero there remains a *residual strain* represented by OC; this is an example of a *plastic deformation.* On reloading the path will, closely, retrace CB and eventually continue the curve OAB.

It is clear from Fig. 10.1 that for this material there is, in general, no unique relation between the stress and the strain, so the theory of elasticity is inappropriate. The discrepancy cannot be explained as a viscous effect, because the behaviour is (except at very high rates of strain) almost independent of the speed at which the deformation is performed. Figure 10.1 also suggests that two phenomena are involved, one being essentially elasticity and involving deformations which vanish on unloading, and the other, called plasticity, giving rise to rate-independent permanent deformations. This idea is supported by the description of the phenomena on the microscopic scale. Materials which exhibit this kind of behaviour are usually crystalline solids. Elastic deformation on the microscopic scale is explained as small recoverable displacements of the atoms which form the crystal lattice from their equilibrium positions. Plastic deformation is caused mainly by permanent slip of neighbouring planes of atoms relative to each other.

To formulate a three-dimensional theory of plasticity we require:

(a) a *yield condition*, which decides whether an element of material is behaving elastically or plastically at a given time;
(b) stress–strain relations for elastic behaviour;
(c) stress–strain relations for plastic behaviour.

Yield condition. This is an inequality of the form

$$f(T_{ij}) \leq k^2 \tag{10.35}$$

where $f(T_{ij})$ is the *yield function*, and k is a parameter which, in general, depends on the deformation history. If $f(T_{ij}) < k^2$, then the material behaves elastically; if $f(T_{ij}) = k^2$ then plastic deformation may occur. The equation $f(T_{ij}) = k^2$ can be regarded as representing a surface (the *yield surface*) in the six-dimensional space of the stress components T_{ij}. Plastic stress states lie on this surface, elastic states in its interior, and stress states outside the yield surface are not attainable for the current value of k.

Any material symmetry restricts the form of $f(T_{ij})$. For example, for an isotropic material the yield function must be expressible as a function of the stress invariants J_1, J_2 and J_3.

For many materials, particularly metals, it is found that to a good approximation the yielding of the material is not affected by a superposed hydrostatic stress. The components S_{ij} of the stress deviator tensor (Section 5.7) are independent of the hydrostatic part of the stress, and for these materials (10.35) may be replaced by

$$f(S_{ij}) \leqslant k^2 \tag{10.36}$$

In the case of an isotropic material, the yield function may now be expressed as a function of the two invariants J_2' and J_3' of $\boldsymbol{S}$.

Elastic stress–strain relations. Before any plastic deformation has occurred, as for example on the section OA of the stress–strain curve in Fig. 10.1, the usual elastic relations apply; for example, for small deformations of an isotropic material we have equations (8.22),

$$T_{ij} = \lambda\delta_{ij}E_{kk} + 2\mu E_{ij} \tag{10.37}$$

For small elastic deformations following a plastic deformation, the relation between $\boldsymbol{T}$ and $\boldsymbol{E}$ is again linear, but the state of zero stress does not correspond to one of zero strain. Thus, for an isotropic material,

$$T_{ij} = \lambda\delta_{ij}(E_{kk} - E_{kk}^{(0)}) + 2\mu(E_{ij} - E_{ij}^{(0)}), \tag{10.38}$$

where $E_{ij}^{(0)}$ represents the residual strain which would result from unloading to zero stress, and which depends on the previous deformation history. The introduction of $E_{ij}^{(0)}$ can be avoided by expressing the elastic stress–strain relation in terms of stress and strain increments or stress and strain rates. Thus (10.37) and (10.38) can be replaced by

$$\delta T_{ij} = \lambda\delta_{ij}\,\delta E_{kk} + 2\mu\,\delta E_{ij} \tag{10.39}$$

or by

$$\dot{T}_{ij} = \lambda\delta_{ij}\dot{E}_{kk} + 2\mu\dot{E}_{ij} = \lambda\delta_{ij}D_{kk} + 2\mu D_{ij} \tag{10.40}$$

where the superposed dot denotes an appropriate time derivative. For finite deformations, these time derivatives are not unambiguous and they require careful definition. No such difficulty arises if attention is restricted to infinitesimal deformations. The inverse of (10.40) is

$$D_{ij} = \frac{1}{2\mu}\dot{T}_{ij} - \frac{\lambda}{2\mu(3\lambda + 2\mu)}\dot{T}_{kk}\delta_{ij} \tag{10.41}$$

Plastic stress–strain relations. The formulation of these is more difficult and controversial, and will not be pursued in detail. The classical approach is to assume that the rate of deformation can be decomposed into an elastic part $D_{ij}^{(e)}$ and a plastic part $D_{ij}^{(p)}$:

$$D_{ij} = D_{ij}^{(e)} + D_{ij}^{(p)} \tag{10.42}$$

The elastic part is related to the stress rate $\dot{T}_{ij}$ by (10.41). For the plastic part the simplest theory postulates (with some justification) that the yield function serves as a plastic potential, in the sense that

$$D_{ij}^{(p)} = \dot{\Lambda}\, \partial f / \partial T_{ij} \tag{10.43}$$

where $\dot{\Lambda}$ is a scalar factor of proportionality which depends on the deformation history. Then, by combining (10.41) and (10.43), we obtain the complete stress–strain relations for an isotropic plastic material, namely

$$D_{ij} = \frac{1}{2\mu} \dot{T}_{ij} - \frac{\lambda}{2\mu(3\lambda + 2\mu)} \dot{T}_{kk} \delta_{ij} + \dot{\Lambda} \frac{\partial f(T_{ij})}{\partial T_{ij}} \tag{10.44}$$

where f can be expressed as a function of J_2' and J_3'.

10.6 Problems

1. The unit cube $0 \leq X_1 \leq 1$, $0 \leq X_2 \leq 1$, $0 \leq X_3 \leq 1$ of incompressible isotropic elastic material undergoes the deformation $x_1 = \lambda X_1 + \alpha X_2$, $x_2 = \lambda^{-1} X_2$, $x_3 = X_3$, where λ and α are constants. Sketch the deformed cube, noting the lengths of its edges. Find the stress, and show that p can be chosen so that no forces act on the surfaces $X_3 = 0$ and $X_3 = 1$. Find the force which must be applied to the face initially given by $X_2 = 1$ to maintain the deformation. Determine the normal in the deformed configuration to the face $X_1 = 1$, and the traction which must be applied to this face to maintain the deformation.

2. A unit cube of incompressible isotropic elastic material undergoes the finite deformation

$$x_1 = \lambda X_1, \qquad x_2 = \lambda^{-1} X_2, \qquad x_3 = X_3$$

where λ is constant. The strain-energy function is

$$W = C_1(I_1 - 3) + C_2(I_2 - 3)$$

where C_1 and C_2 are constants. Sketch the deformed cube, noting the lengths of its edges. Find the stress and hence determine the total loads F_1, F_2 and F_3 acting on the faces normal to the X_1, X_2 and X_3 directions. Show that when $C_1 > 3C_2 > 0$, there are three values of λ for which the body is in equilibrium with $F_1 = F_2 = F_3$, and find these values.

3. Show that the constitutive equation for an elastic solid can be expressed in the form

$$T_{ij} = \frac{1}{2}\frac{\rho}{\rho_0}\frac{\partial x_i}{\partial X_R}\frac{\partial x_j}{\partial X_S}\left(\frac{\partial W}{\partial \gamma_{RS}} + \frac{\partial W}{\partial \gamma_{SR}}\right)$$

4. For a particular transversely isotropic elastic solid, with preferred direction that of the X_1-axis, W has the form

$$W = \alpha C_{PP}C_{QQ} + \beta C_{PQ}C_{PQ} + \gamma C_{11}^2 + \delta(C_{12}^2 + C_{13}^2)$$

where α, β, γ and δ are constants. Find the constitutive equation for $\boldsymbol{T}$, and hence find the stress in a body of this material subjected to the uniform expansion

$$x_1 = \lambda X_1, \qquad x_2 = \lambda X_2, \qquad x_3 = \lambda X_3$$

5. Suppose that the stress in a solid is given by a relation of the form $\boldsymbol{T} = \boldsymbol{\chi}(\boldsymbol{F})$. Show that if the stress is independent of rotations of the deformed body, then $\boldsymbol{\chi}$ must satisfy the relation $\boldsymbol{\chi}(\boldsymbol{M}\cdot\boldsymbol{F}) = \boldsymbol{M}\cdot\boldsymbol{\chi}(\boldsymbol{F})\cdot\boldsymbol{M}^{\mathrm{T}}$ for all proper orthogonal tensors $\boldsymbol{M}$. Verify that a sufficient condition for this relation to be satisfied is that $\boldsymbol{\chi}$ can be expressed in the form $\boldsymbol{\chi} = \boldsymbol{F}\cdot\boldsymbol{\Psi}(\boldsymbol{C})\cdot\boldsymbol{F}^{\mathrm{T}}$. Use the representation theorem given in the Appendix to obtain the most general such form for $\boldsymbol{\chi}$ in the case in which the material is isotropic.

6. Derive the constitutive equation $\boldsymbol{T} = -p\boldsymbol{I} + 2\mu\boldsymbol{E}$, for incompressible isotropic linear elasticity, as a first approximation, for small displacement gradients, to equation (10.20).

7. Show that the most general incompressible Reiner–Rivlin fluid (10.32) for which the stress components are quadratic functions of the components D_{ij} has the constitutive equation $\boldsymbol{T} = -p\boldsymbol{I} + \alpha_0\boldsymbol{D} + \beta_0\boldsymbol{D}^2$, where α_0 and β_0 are constants.

8. Show that a velocity field $v_1 = v(x_2)$, $v_2 = 0$, $v_3 = 0$, is a possible flow in every incompressible Reiner–Rivlin fluid (10.32). If this flow takes place between infinite parallel plates at $x_2 = \pm d$, determine the pressure gradient (that is, $-\partial T_{11}/\partial x_1$) required to

maintain this flow and the tangential forces acting on unit area of each of the plates.

9. The stress in a certain Reiner–Rivlin fluid is given by $\boldsymbol{T} = -p\boldsymbol{I} + \mu(1 + \alpha \operatorname{tr} \boldsymbol{D}^2)\boldsymbol{D} + \beta\boldsymbol{D}^2$, where α, β and μ are constants. Determine the stress in the fluid arising from the velocity field $v_1 = -x_2\omega(x_3)$, $v_2 = x_1\omega(x_3)$, $v_3 = 0$. Show that if $\omega = Ax_3 + B$, where A and B are constants, the equations of motion are satisfied only if $A = 0$ or if the acceleration terms can be neglected. In the latter case, find values of A and B corresponding to flow between parallel plates at $x_3 = 0$ and $x_3 = h$, the former being at rest and the latter rotating about the x_3-axis with angular speed Ω.

10. The behaviour of certain viscous fluids is often modelled by the constitutive equation

$$T_{ij} = -p\delta_{ij} + 2\mu(K_2)D_{ij}, \text{ where } K_2 = 2D_{ij}D_{ij},\ \mu(K_2) = kK_2^{(n-1)/2}$$

and k and n are positive constants (and $n = 1$ corresponds to a Newtonian fluid). Such a power-law fluid undergoes simple shearing flow between two large parallel plates a distance h apart, such that one plate is held fixed and the other moves with constant speed U in its plane. Find the shearing force per unit area on the plates and the apparent viscosity μ as a function of the shear rate U/h.

11. The constitutive equation $\boldsymbol{T} = -p\boldsymbol{I} + 2\mu_0(2 \operatorname{tr} \boldsymbol{D}^2)^{\alpha}\boldsymbol{D}$, where μ_0 and α are constants, models a class of Reiner–Rivlin fluids. Show that these fluids can undergo the steady rectilinear shear flow $v_1 = v(x_2)$, $v_2 = 0$, $v_3 = 0$, provided $p = p_0 + kx_1$, where p_0 and k are constants.

12. Determine the tensors $\boldsymbol{C}(\tau)$ and $\boldsymbol{A}^{(n)}(t)$ which are defined in Problem 5 of Chapter 9 for the motion $x_1(\tau) = X_1$, $x_2(\tau) = X_2$, $x_3(\tau) = X_3 + \gamma\tau \tan^{-1}(X_2/X_1)$, where γ is a constant. The stress in a fluid is given by $\boldsymbol{T} = -p\boldsymbol{I} + \mu\boldsymbol{A}^{(1)} + \sigma\boldsymbol{A}^{(2)}$, where μ and σ are functions of $\operatorname{tr} \boldsymbol{A}^{(2)}$ and p is arbitrary. Show that if p is a function of r only ($r^2 = x_1^2 + x_2^2$), then the equations of motion are satisfied provided that

$$\frac{dp}{dr} = -\frac{2\gamma^2\sigma}{r^3}$$

13. The stress in a particular incompressible non-Newtonian fluid is given by $\boldsymbol{T} = -p\boldsymbol{I} + \mu \int_0^\infty \exp(-ks)\{\boldsymbol{C}(\tau) - \boldsymbol{I}\}\, ds$, where $s = t - \tau$

and $\boldsymbol{C}(\tau)$ is defined in Problem 5 of Chapter 9. Determine the stress in the fluid due to the displacement field

$$x_1(\tau)=x_1-f(x_2)(\cos\omega t-\cos\omega\tau)-g(x_2)(\sin\omega t-\sin\omega\tau)$$

$$x_2(\tau)=x_2,\qquad x_3(\tau)=x_3$$

if df/dx_2 and dg/dx_2 are small enough for their squares to be neglected.

11

Cylindrical and spherical polar coordinates

11.1 Curvilinear coordinates

So far we have used only rectangular cartesian coordinates, and this is the simplest way to formulate the general equations of continuum mechanics and the constitutive equations of various ideal materials. However, for the solution of particular problems, it is often preferable to work in terms of other systems of coordinates. In particular, it is usually desirable to use cylindrical polar coordinates for configurations which have an element of symmetry about an axis, and to use spherical polar coordinates when there is some symmetry about a point. It is therefore useful to express the main equations in terms of these other coordinate systems.

It is possible to develop elegantly the equations of continuum mechanics in terms of general curvilinear coordinates. Results in any particular coordinate system can then be obtained by making the appropriate specializations. However, this procedure requires extensive use of general curvilinear tensor analysis, which we prefer to avoid in this introductory text. Also, it is only very rarely that coordinate systems other than rectangular cartesian, cylindrical polar and spherical polar coordinates can be employed profitably. Accordingly, we shall derive directly some results in cylindrical and spherical polars, even though these results could be obtained more concisely by the use of general tensor analysis.

11.2 Cylindrical polar coordinates

Cylindrical polar coordinates r, ϕ, z $(0 \leqslant \phi < 2\pi)$ are related to cartesian coordinates x_1, x_2, x_3 by

$$x_1 = r\cos\phi, \qquad x_2 = r\sin\phi, \qquad x_3 = z \tag{11.1}$$

$$r = (x_1^2 + x_2^2)^{\frac{1}{2}}, \qquad \phi = \tan^{-1}(x_2/x_1), \qquad z = x_3 \tag{11.2}$$

The base vectors of the r, ϕ, z coordinate system are unit vectors

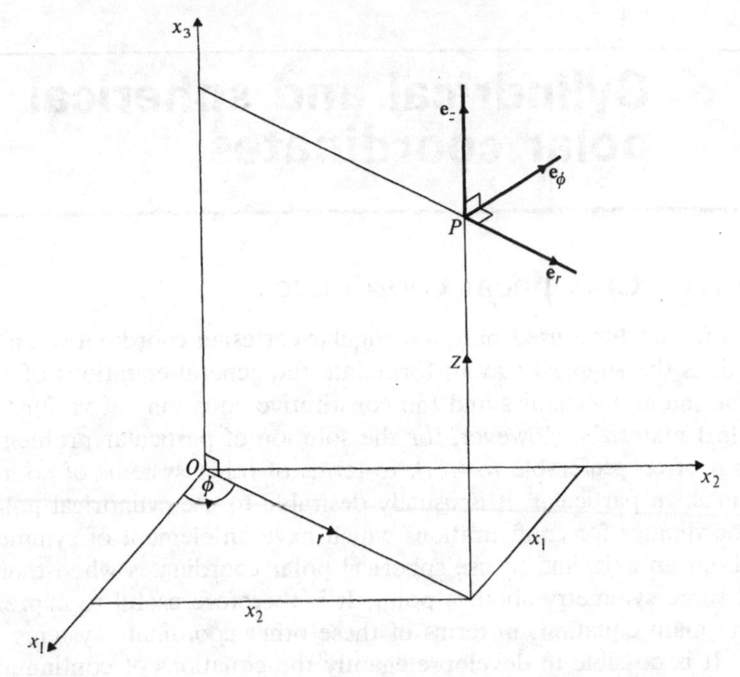

Figure 11.1 Base vectors for cylindrical polar coordinates

directed in the radial, tangential and axial directions, as illustrated in Fig. 11.1. They are denoted by $\boldsymbol{e}_r$, $\boldsymbol{e}_\phi$ and $\boldsymbol{e}_z$, and they are mutually orthogonal. Thus

$$\begin{aligned} \boldsymbol{e}_r &= \boldsymbol{e}_1 \cos\phi + \boldsymbol{e}_2 \sin\phi, \\ \boldsymbol{e}_\phi &= -\boldsymbol{e}_1 \sin\phi + \boldsymbol{e}_2 \cos\phi, \\ \boldsymbol{e}_z &= \boldsymbol{e}_3 \end{aligned} \tag{11.3}$$

$$\begin{aligned} \boldsymbol{e}_1 &= \boldsymbol{e}_r \cos\phi - \boldsymbol{e}_\phi \sin\phi, \\ \boldsymbol{e}_2 &= \boldsymbol{e}_r \sin\phi + \boldsymbol{e}_\phi \cos\phi, \\ \boldsymbol{e}_3 &= \boldsymbol{e}_z \end{aligned} \tag{11.4}$$

We define the matrix $\mathbf{R}$ to be

$$\mathbf{R} = \begin{pmatrix} \cos\phi & \sin\phi & 0 \\ -\sin\phi & \cos\phi & 0 \\ 0 & 0 & 1 \end{pmatrix} \tag{11.5}$$

and then (11.3) and (11.4) may be written as

$$(\boldsymbol{e}_r \quad \boldsymbol{e}_\phi \quad \boldsymbol{e}_z)^{\mathrm{T}} = \mathbf{R}(\boldsymbol{e}_1 \quad \boldsymbol{e}_2 \quad \boldsymbol{e}_3)^{\mathrm{T}}, \qquad (\boldsymbol{e}_1 \quad \boldsymbol{e}_2 \quad \boldsymbol{e}_3)^{\mathrm{T}} = \mathbf{R}^{\mathrm{T}}(\boldsymbol{e}_r \quad \boldsymbol{e}_\phi \quad \boldsymbol{e}_z)^{\mathrm{T}} \tag{11.6}$$

It is easily verified that $\mathbf{R}$ is an orthogonal matrix.

Suppose a vector $\boldsymbol{a}$ has components a_i in the coordinate system x_i and components a_r, a_ϕ, a_z in the system r, ϕ, z, so that

$$\boldsymbol{a} = a_i \boldsymbol{e}_i = a_r \boldsymbol{e}_r + a_\phi \boldsymbol{e}_\phi + a_z \boldsymbol{e}_z \tag{11.7}$$

Let

$$\mathbf{a} = (a_1 \quad a_2 \quad a_3)^{\mathrm{T}}, \qquad \mathbf{a}^* = (a_r \quad a_\phi \quad a_z)^{\mathrm{T}} \tag{11.8}$$

be the column matrices formed from the components of $\boldsymbol{a}$ in the two coordinate systems. Then from (11.4) and (11.7)

$$\mathbf{a}^* = \mathbf{R}\mathbf{a}, \qquad \mathbf{a} = \mathbf{R}^{\mathrm{T}}\mathbf{a}^* \tag{11.9}$$

A second-order tensor $\mathbf{A} = A_{ij}\boldsymbol{e}_i \otimes \boldsymbol{e}_j$ can be written as

$$\begin{aligned}\mathbf{A} = {} & A_{rr}\boldsymbol{e}_r \otimes \boldsymbol{e}_r + A_{r\phi}\boldsymbol{e}_r \otimes \boldsymbol{e}_\phi + A_{rz}\boldsymbol{e}_r \otimes \boldsymbol{e}_z \\ & + A_{\phi r}\boldsymbol{e}_\phi \otimes \boldsymbol{e}_r + A_{\phi\phi}\boldsymbol{e}_\phi \otimes \boldsymbol{e}_\phi + A_{\phi z}\boldsymbol{e}_\phi \otimes \boldsymbol{e}_z \\ & + A_{zr}\boldsymbol{e}_z \otimes \boldsymbol{e}_r + A_{z\phi}\boldsymbol{e}_z \otimes \boldsymbol{e}_\phi + A_{zz}\boldsymbol{e}_z \otimes \boldsymbol{e}_z\end{aligned}$$

or, more concisely, in matrix notation as

$$\mathbf{A} = (\boldsymbol{e}_r \quad \boldsymbol{e}_\phi \quad \boldsymbol{e}_z) \otimes \mathbf{A}^*(\boldsymbol{e}_r \quad \boldsymbol{e}_\phi \quad \boldsymbol{e}_z)^{\mathrm{T}} \tag{11.10}$$

where

$$\mathbf{A}^* = \begin{pmatrix} A_{rr} & A_{r\phi} & A_{rz} \\ A_{\phi r} & A_{\phi\phi} & A_{\phi z} \\ A_{zr} & A_{z\phi} & A_{zz} \end{pmatrix} \tag{11.11}$$

is the matrix of components of $\mathbf{A}$ referred to r, ϕ, z coordinates. From (11.6) and (11.10) there follow

$$\mathbf{A} = \mathbf{R}^{\mathrm{T}}\mathbf{A}^*\mathbf{R}, \qquad \text{and} \qquad \mathbf{A}^* = \mathbf{R}\mathbf{A}\mathbf{R}^{\mathrm{T}} \tag{11.12}$$

where $\mathbf{A} = (A_{ij})$ is the matrix of components of $\mathbf{A}$ referred to x_i coordinates. From (11.12), it follows that if $\mathbf{A}$ is a symmetric matrix, then so is $\mathbf{A}^*$, and if $\mathbf{A}$ is an anti-symmetric matrix, then so is $\mathbf{A}^*$. Since $\mathbf{R}$ is orthogonal, the eigenvalues of $\mathbf{A}$ and $\mathbf{A}^*$ are the same, so the principal values of $\mathbf{A}$ are the roots of the equation

$$\det(\mathbf{A}^* - A\mathbf{I}) = 0$$

Moreover, the invariants I_1, I_2 and I_3 of $\mathbf{A}$ may be written as

$$I_1 = \operatorname{tr}\mathbf{A}^*, \qquad I_2 = \tfrac{1}{2}\{(\operatorname{tr}\mathbf{A}^*)^2 - \operatorname{tr}\mathbf{A}^{*2}\}, \qquad I_3 = \det\mathbf{A}^* \tag{11.13}$$

Referred to cylindrical polar coordinates, the gradient of a scalar $\psi(r, \phi, z)$ and the divergence of a vector $\boldsymbol{a}(r, \phi, z)$ are, respectively

$$\begin{aligned} \operatorname{grad} \psi = \boldsymbol{\nabla}\psi &= \frac{\partial \psi}{\partial r}\boldsymbol{e}_r + \frac{1}{r}\frac{\partial \psi}{\partial \phi}\boldsymbol{e}_\phi + \frac{\partial \psi}{\partial z}\boldsymbol{e}_z \\ \operatorname{div} \mathbf{A} = \boldsymbol{\nabla}\cdot\boldsymbol{a} &= \frac{1}{r}\frac{\partial (ra_r)}{\partial r} + \frac{1}{r}\frac{\partial a_\phi}{\partial \phi} + \frac{\partial a_z}{\partial z} \end{aligned} \tag{11.14}$$

The material derivative of $\psi(r, \phi, z, t)$ is then given by (4.18) as

$$\dot{\psi} = \frac{\partial \psi}{\partial t} + v_r\frac{\partial \psi}{\partial r} + \frac{v_\phi}{r}\frac{\partial \psi}{\partial \phi} + v_z\frac{\partial \psi}{\partial z} \tag{11.15}$$

If $\boldsymbol{v} = v_r\boldsymbol{e}_r + v_\phi\boldsymbol{e}_\phi + v_z\boldsymbol{e}_z$ is the velocity vector, then from (4.23) the acceleration vector $\boldsymbol{f}$ is given by

$$\frac{\partial \boldsymbol{v}}{\partial t} + \left(v_r\frac{\partial}{\partial r} + \frac{v_\phi}{r}\frac{\partial}{\partial \phi} + v_z\frac{\partial}{\partial z}\right)(v_r\boldsymbol{e}_r + v_\phi\boldsymbol{e}_\phi + v_z\boldsymbol{e}_z) \tag{11.16}$$

Suppose that the matrix of components of the stress tensor referred to r, ϕ, z coordinates is $\mathbf{T}^*$, where

$$\mathbf{T}^* = \begin{pmatrix} T_{rr} & T_{r\phi} & T_{rz} \\ T_{\phi r} & T_{\phi\phi} & T_{\phi z} \\ T_{zr} & T_{z\phi} & T_{zz} \end{pmatrix} \tag{11.17}$$

and that $\mathbf{T} = (T_{ij})$. Then

$$\mathbf{T} = \mathbf{R}^{\mathrm{T}}\mathbf{T}^*\mathbf{R}, \qquad \mathbf{T}^* = \mathbf{R}\mathbf{T}\mathbf{R}^{\mathrm{T}} \tag{11.18}$$

Because (11.18) are important relations, we give them in full, as follows:

$$\left.\begin{aligned} T_{11} &= T_{rr}\cos^2\phi - 2T_{r\phi}\cos\phi\sin\phi + T_{\phi\phi}\sin^2\phi \\ T_{22} &= T_{rr}\sin^2\phi + 2T_{r\phi}\cos\phi\sin\phi + T_{\phi\phi}\cos^2\phi \\ T_{12} &= (T_{rr} - T_{\phi\phi})\cos\phi\sin\phi + T_{r\phi}(\cos^2\phi - \sin^2\phi) \\ T_{13} &= T_{rz}\cos\phi - T_{\phi z}\sin\phi \\ T_{23} &= T_{rz}\sin\phi + T_{\phi z}\cos\phi, \qquad T_{33} = T_{zz} \\ T_{rr} &= T_{11}\cos^2\phi + 2T_{12}\cos\phi\sin\phi + T_{22}\sin^2\phi \\ T_{\phi\phi} &= T_{11}\sin^2\phi - 2T_{12}\cos\phi\sin\phi + T_{22}\cos^2\phi \\ T_{r\phi} &= -(T_{11} - T_{22})\cos\phi\sin\phi + T_{12}(\cos^2\phi - \sin^2\phi) \\ T_{rz} &= T_{13}\cos\phi + T_{23}\sin\phi, \\ T_{\phi z} &= -T_{13}\sin\phi + T_{23}\cos\phi, \qquad T_{zz} = T_{33} \end{aligned}\right\} \tag{11.19}$$

Let a surface have normal $\boldsymbol{n}$, where

$$\boldsymbol{n} = n_i \boldsymbol{e}_i = n_r \boldsymbol{e}_r + n_\phi \boldsymbol{e}_\phi + n_z \boldsymbol{e}_z \tag{11.20}$$

Then by (5.9), the traction vector on the surface is $n_i T_{ij} \boldsymbol{e}_j$, and using (11.18) and (11.20) this can be expressed as

$$(n_r \quad n_\phi \quad n_z)\mathbf{T}^*(\boldsymbol{e}_r \quad \boldsymbol{e}_\phi \quad \boldsymbol{e}_z)^{\mathrm{T}}$$

From (5.27) and (11.18), the stress invariants J_1, J_2 and J_3 can be written in the forms

$$J_1 = \operatorname{tr}\mathbf{T}^*, \qquad J_2 = \tfrac{1}{2}\{\operatorname{tr}\mathbf{T}^{*2} - (\operatorname{tr}\mathbf{T}^*)^2\}, \qquad J_3 = \det \mathbf{T}^* \tag{11.21}$$

Now consider a finite deformation in which a typical particle which in the reference configuration has cylindrical polar coordinates R, Φ, Z moves to the position with cylindrical polar coordinates r, ϕ, z, where

$$X_1 = R\cos\Phi, \qquad X_2 = R\sin\Phi, \qquad X_3 = Z \tag{11.22}$$

$$x_1 = r\cos\phi \qquad x_2 = r\sin\phi, \qquad x_3 = z \tag{11.23}$$

The motion can be described by equations of the form

$$r = r(R, \Phi, Z), \qquad \phi = \phi(R, \Phi, Z) \qquad z = z(R, \Phi, Z) \tag{11.24}$$

Let

$$\mathbf{F} = (F_{iR}) = (\partial x_i/\partial X_R) \tag{11.25}$$

and, in addition to the matrix $\mathbf{R}$ defined by (11.5), introduce an orthogonal matrix $\mathbf{R}_0$, where

$$\mathbf{R}_0 = \begin{pmatrix} \cos\Phi & \sin\Phi & 0 \\ -\sin\Phi & \cos\Phi & 0 \\ 0 & 0 & 1 \end{pmatrix} \tag{11.26}$$

We also observe, from (11.22), that

$$\begin{aligned} \frac{\partial}{\partial R} &= \cos\Phi\frac{\partial}{\partial X_1} + \sin\Phi\frac{\partial}{\partial X_2}, \\ \frac{1}{R}\frac{\partial}{\partial\Phi} &= -\sin\Phi\frac{\partial}{\partial X_1} + \cos\Phi\frac{\partial}{\partial X_2}, \qquad \frac{\partial}{\partial Z} = \frac{\partial}{\partial X_3} \end{aligned} \tag{11.27}$$

Then it can be shown from (11.5), (11.23), (11.25), (11.26) and

(11.27), after a little manipulation, that

$$\mathbf{F}^* = \mathbf{R}\mathbf{F}\mathbf{R}_0^{\mathrm{T}} = \begin{pmatrix} \dfrac{\partial r}{\partial R} & \dfrac{1}{R}\dfrac{\partial r}{\partial \Phi} & \dfrac{\partial r}{\partial Z} \\ r\dfrac{\partial \phi}{\partial R} & \dfrac{r}{R}\dfrac{\partial \phi}{\partial \Phi} & r\dfrac{\partial \phi}{\partial Z} \\ \dfrac{\partial z}{\partial R} & \dfrac{1}{R}\dfrac{\partial z}{\partial \Phi} & \dfrac{\partial z}{\partial Z} \end{pmatrix} \qquad (11.28)$$

Suppose that $\mathbf{B} = (B_{ij}) = \mathbf{F}\mathbf{F}^{\mathrm{T}}$ is the matrix of components of $\boldsymbol{B}$, referred to x_i coordinates, and let $\mathbf{B}^*$ be the matrix of components of $\boldsymbol{B}$ referred to r, ϕ, z coordinates. Then

$$\mathbf{B}^* = \mathbf{R}\mathbf{B}\mathbf{R}^{\mathrm{T}} = \mathbf{R}\mathbf{F}\mathbf{F}^{\mathrm{T}}\mathbf{R}^{\mathrm{T}} = \mathbf{R}\mathbf{F}\mathbf{R}_0^{\mathrm{T}}\mathbf{R}_0\mathbf{F}^{\mathrm{T}}\mathbf{R}^{\mathrm{T}} = \mathbf{F}^*\mathbf{F}^{*\mathrm{T}} \qquad (11.29)$$

Hence $\mathbf{B}^*$ is readily calculated from (11.24) and (11.28). Similarly, if $\mathbf{C} = (C_{RS}) = \mathbf{F}^{\mathrm{T}}\mathbf{F}$ is the matrix of components of $\boldsymbol{C}$ referred to X_R coordinates, and $\mathbf{C}^*$ is the matrix of components of $\boldsymbol{C}$ referred to R, Φ, Z coordinates, then

$$\mathbf{C}^* = \mathbf{R}_0\mathbf{C}\mathbf{R}_0^{\mathrm{T}} = \mathbf{R}_0\mathbf{F}^{\mathrm{T}}\mathbf{F}\mathbf{R}_0^{\mathrm{T}} = \mathbf{R}_0\mathbf{F}^{\mathrm{T}}\mathbf{R}^{\mathrm{T}}\mathbf{R}\mathbf{F}\mathbf{R}_0^{\mathrm{T}} = \mathbf{F}^{*\mathrm{T}}\mathbf{F}^* \qquad (11.30)$$

We also note, for future reference, that

$$\mathbf{R}\mathbf{B}^{-1}\mathbf{R}^{\mathrm{T}} = \mathbf{R}(\mathbf{F}^{-1})^{\mathrm{T}}\mathbf{R}_0^{\mathrm{T}}\mathbf{R}_0\mathbf{F}^{-1}\mathbf{R}^{\mathrm{T}} = (\mathbf{F}^{*\mathrm{T}})^{-1}(\mathbf{F}^*)^{-1} = (\mathbf{B}^*)^{-1} \qquad (11.31)$$

For a *small* displacement $\boldsymbol{u} = u_r\boldsymbol{e}_r + u_\phi\boldsymbol{e}_\phi + u_z\boldsymbol{e}_z$, we have

$u_r = u_1 \cos\phi + u_2 \sin\phi$, $u_\phi = -u_1 \sin\phi + u_2 \cos\phi$, $u_z = u_3$

Then $\partial u_i/\partial x_j \simeq \partial u_i/\partial X_j$, and it follows from (6.26) and (11.5) that

$$\hat{\mathbf{F}} - \mathbf{I} = \mathbf{R}\mathbf{F}\mathbf{R}^{\mathrm{T}} - \mathbf{I} \simeq \begin{pmatrix} \dfrac{\partial u_r}{\partial r} & \dfrac{1}{r}\left(\dfrac{\partial u_r}{\partial \phi} - u_\phi\right) & \dfrac{\partial u_r}{\partial z} \\ \dfrac{\partial u_\phi}{\partial r} & \dfrac{1}{r}\left(u_r + \dfrac{\partial u_\phi}{\partial \phi}\right) & \dfrac{\partial u_\phi}{\partial z} \\ \dfrac{\partial u_z}{\partial r} & \dfrac{1\partial u_z}{r\partial \phi} & \dfrac{\partial u_z}{\partial z} \end{pmatrix} \qquad (11.32)$$

and, in the small-displacement approximation, there is no need to distinguish between R, Φ, Z and r, ϕ, z in (11.32). The matrix $\mathbf{E}^*$ of infinitesimal strain components, and the matrix $\boldsymbol{\Omega}^*$ of infinitesimal rotation components, referred to cylindrical polar coordinates, are then given by

$$\mathbf{E}^* = \tfrac{1}{2}(\hat{\mathbf{F}} + \hat{\mathbf{F}}^{\mathrm{T}}) - \mathbf{I}, \qquad \Omega^* = \tfrac{1}{2}(\hat{\mathbf{F}} - \hat{\mathbf{F}}^{\mathrm{T}}) \qquad (11.33)$$

Similarly the matrix $\mathbf{L}^*$ of the components of the velocity gradient tensor $\boldsymbol{L}$, referred to coordinates (r, ϕ, z), is

$$\boldsymbol{L}^* = \begin{pmatrix} \dfrac{\partial v_r}{\partial r} & \dfrac{1}{r}\left(\dfrac{\partial v_r}{\partial \phi} - v_\phi\right) & \dfrac{\partial v_r}{\partial z} \\ \dfrac{\partial v_\phi}{\partial r} & \dfrac{1}{r}\left(v_r + \dfrac{\partial v_\phi}{\partial \phi}\right) & \dfrac{\partial v_\phi}{\partial z} \\ \dfrac{\partial v_z}{\partial r} & \dfrac{1}{r}\dfrac{\partial v_z}{\partial \phi} & \dfrac{\partial v_z}{\partial z} \end{pmatrix} \tag{11.34}$$

The expression (11.34) is exact. The matrices $\mathbf{D}^*$ and $\mathbf{W}^*$ of the components, referred to r, ϕ, z coordinates, of the rate-of-deformation tensor $\boldsymbol{D}$ and the vorticity tensor $\boldsymbol{W}$ are then given by

$$\mathbf{D}^* = \tfrac{1}{2}(\mathbf{L}^* + \mathbf{L}^{*\mathrm{T}}), \qquad \mathbf{W}^* = \tfrac{1}{2}(\mathbf{L}^* - \mathbf{L}^{*\mathrm{T}}) \tag{11.35}$$

From (11.28) we have $\det \mathbf{F} = \det \mathbf{F}^*$. Hence from (7.8),

$$\frac{\rho_0}{\rho} = \frac{\mathrm{d}v}{\mathrm{d}V} = \det \mathbf{F}^* \tag{11.36}$$

and in an incompressible material, $\det \mathbf{F}^* = 1$. The Eulerian form of the mass-conservation equation is given by (7.11), and can be expressed in terms of the components of $\boldsymbol{v}$ referred to cylindrical polar coordinates by expressing $\operatorname{div}(\rho\boldsymbol{v})$ in these coordinates.

The equation of motion (7.22) can be expressed in terms of cylindrical polar coordinates by resolving the body force and acceleration into components referred to these coordinates. Let (b_r, b_ϕ, b_z) be components of $\boldsymbol{b}$, and let (f_r, f_ϕ, f_z) be components of $\boldsymbol{f}$, in cylindrical polar coordinates. Then, from (7.22) and (11.9)

$$\rho\begin{pmatrix} f_r - b_r \\ f_\phi - b_\phi \\ f_z - b_z \end{pmatrix} = \rho\begin{pmatrix} \cos\phi & \sin\phi & 0 \\ -\sin\phi & \cos\phi & 0 \\ 0 & 0 & 1 \end{pmatrix}\begin{pmatrix} f_1 - b_1 \\ f_2 - b_2 \\ f_3 - b_3 \end{pmatrix}$$

$$= \begin{pmatrix} \cos\phi & \sin\phi & 0 \\ -\sin\phi & \cos\phi & 0 \\ 0 & 0 & 1 \end{pmatrix}\begin{pmatrix} \dfrac{\partial T_{11}}{\partial x_1} + \dfrac{\partial T_{21}}{\partial x_2} + \dfrac{\partial T_{31}}{\partial x_3} \\ \dfrac{\partial T_{12}}{\partial x_1} + \dfrac{\partial T_{22}}{\partial x_2} + \dfrac{\partial T_{32}}{\partial x_3} \\ \dfrac{\partial T_{13}}{\partial x_1} + \dfrac{\partial T_{23}}{\partial x_2} + \dfrac{\partial T_{33}}{\partial x_3} \end{pmatrix} \tag{11.37}$$

From (11.1) we have

$$\begin{aligned}\frac{\partial}{\partial x_1}&=\cos\phi\frac{\partial}{\partial r}-r^{-1}\sin\phi\frac{\partial}{\partial \phi}\\ \frac{\partial}{\partial x_2}&=\sin\phi\frac{\partial}{\partial r}+r^{-1}\cos\phi\frac{\partial}{\partial \phi},\qquad \frac{\partial}{\partial x_3}=\frac{\partial}{\partial z}\end{aligned}\tag{11.38}$$

By introducing (11.19) and (11.38) into (11.37) it follows, after some manipulations, that

$$\begin{aligned}\frac{\partial T_{rr}}{\partial r}+\frac{1}{r}\frac{\partial T_{r\phi}}{\partial \phi}+\frac{\partial T_{rz}}{\partial z}+\frac{T_{rr}-T_{\phi\phi}}{r}+\rho b_r&=\rho f_r\\ \frac{\partial T_{r\phi}}{\partial r}+\frac{1}{r}\frac{\partial T_{\phi\phi}}{\partial \phi}+\frac{\partial T_{\phi z}}{\partial z}+\frac{2T_{r\phi}}{r}+\rho b_\phi&=\rho f_\phi\\ \frac{\partial T_{rz}}{\partial r}+\frac{1}{r}\frac{\partial T_{\phi z}}{\partial \phi}+\frac{\partial T_{zz}}{\partial z}+\frac{T_{rz}}{r}+\rho b_z&=\rho f_z\end{aligned}\tag{11.39}$$

Equations (11.39) are the equations of motion referred to r, ϕ, z coordinates. These equations can also be derived by considering the forces acting on an elementary region bounded by the coordinate surfaces

$$r=r_0,\qquad r=r_0+\delta r,\qquad \phi=\phi_0,\qquad \phi=\phi_0+\delta\phi$$

$$z=z_0\quad\text{and}\quad z=z_0+\delta z$$

Constitutive equations are most easily expressed in terms of cylindrical polar coordinates by multiplying the appropriate expression for the matrix $\mathbf{T}=(T_{ij})$ of stress components on the left by $\mathbf{R}$ and on the right by $\mathbf{R}^\mathrm{T}$. For example, for an isotropic linear elastic solid, we obtain from (8.22)

$$\mathbf{RTR}^\mathrm{T}=\lambda\mathbf{RR}^\mathrm{T}\operatorname{tr}\mathbf{E}+2\mu\mathbf{RER}^\mathrm{T}$$

However, $\mathbf{RTR}^\mathrm{T}=\mathbf{T}^*$, $\mathbf{RR}^\mathrm{T}=\mathbf{I}$, $\operatorname{tr}\mathbf{E}=\operatorname{tr}\mathbf{E}^*$ and $\mathbf{RER}^\mathrm{T}=\mathbf{E}^*$, and so

$$\mathbf{T}^*=\lambda\mathbf{I}\operatorname{tr}\mathbf{E}^*+2\mu\mathbf{E}^*\tag{11.40}$$

where λ and μ are elastic constants. Similarly, the constitutive equation (8.25) for a Newtonian viscous fluid can be expressed in the form

$$\mathbf{T}^*=(-p+\lambda\operatorname{tr}\mathbf{D}^*)\mathbf{I}+2\mu\mathbf{D}^*\tag{11.41}$$

where the pressure $-p$ and the viscosity coefficients λ and μ are functions of the density and the temperature.

From (9.27), (9.29), (11.29) and (11.30), it follows that the strain invariants I_1, I_2 and I_3 can be expressed as

$$\begin{aligned} I_1 &= \operatorname{tr}\mathbf{C}^* = \operatorname{tr}\mathbf{B}^* \\ I_2 &= \tfrac{1}{2}\{(\operatorname{tr}\mathbf{C}^*)^2 - \operatorname{tr}\mathbf{C}^{*2}\} = \tfrac{1}{2}\{(\operatorname{tr}\mathbf{B}^*)^2 - \operatorname{tr}\mathbf{B}^{*2}\} \\ &= I_3 \operatorname{tr}\mathbf{C}^{*-1} = I_3 \operatorname{tr}\mathbf{B}^{*-1} \\ I_3 &= \det\mathbf{C}^* = \det\mathbf{B}^* \end{aligned} \tag{11.42}$$

The constitutive equation (10.18) for an isotropic elastic solid gives

$$\mathbf{RTR}^{\mathrm{T}} = 2(I_3)^{-\frac{1}{2}}\{(I_2W_2 + I_3W_3)\mathbf{I} + W_1\mathbf{RBR}^{\mathrm{T}} - I_3W_2\mathbf{RB}^{-1}\mathbf{R}^{\mathrm{T}}\}$$

which, after using (11.18), (11.29) and (11.31), takes the form

$$\mathbf{T}^* = 2(I_3)^{-\frac{1}{2}}\{(I_2W_2 + I_3W_3)\mathbf{I} + W_1\mathbf{B}^* - I_3W_2\mathbf{B}^{*-1}\} \tag{11.43}$$

If the material is also incompressible, this becomes

$$\mathbf{T}^* = -p\mathbf{I} + 2W_1\mathbf{B}^* - 2W_2\mathbf{B}^{*-1} \tag{11.44}$$

In a similar manner, the constitutive equation (10.32) for a Reiner–Rivlin fluid can be expressed in the form

$$\mathbf{T}^* = -p\mathbf{I} + \alpha\mathbf{D}^* + \beta\mathbf{D}^{*2} \tag{11.45}$$

where p, α and β are functions of density, temperature, and of $\operatorname{tr}\mathbf{D}^*$, $\frac{1}{2}\{(\operatorname{tr}\mathbf{D}^*)^2 - \frac{1}{2}\operatorname{tr}\mathbf{D}^{*2}\}$ and $\det\mathbf{D}^*$.

11.3 Spherical polar coordinates

Spherical polar coordinates s, θ, ϕ $(0 \leq \theta \leq \pi, 0 \leq \phi < 2\pi)$ are related to cylindrical polar coordinates r, ϕ, z by

$$r = s\sin\theta, \qquad \phi = \phi, \qquad z = s\cos\theta \tag{11.46}$$

$$s = (r^2 + z^2)^{\frac{1}{2}}, \qquad \theta = \tan^{-1}(r/z), \qquad \phi = \phi \tag{11.47}$$

and to cartesian coordinates x_1, x_2, x_3 by

$$x_1 = s\sin\theta\cos\phi, \qquad x_2 = s\sin\theta\sin\phi, \qquad x_3 = s\cos\theta \tag{11.48}$$

$$\begin{aligned} &s = (x_1^2 + x_2^2 + x_3^2)^{\frac{1}{2}}, \qquad \theta = \tan^{-1}\{(x_1^2 + x_2^2)^{\frac{1}{2}}/x_3\} \\ &\phi = \tan^{-1}(x_2/x_1) \end{aligned} \tag{11.49}$$

Vector and tensor equations can be expressed in terms of spherical polar coordinates in a similar manner to that which was employed in Section 11.2 for cylindrical polar coordinates, although the algebra involved is slightly more complicated. It is often convenient to employ cylindrical polar coordinates as an intermediate stage between cartesian coordinates and spherical polar coordinates. As the approach is analogous to that of Section 11.2, we omit some details of the derivations of the results presented below.

The base vectors of the s, θ, ϕ system are denoted $\boldsymbol{e}_s$, $\boldsymbol{e}_\theta$ and $\boldsymbol{e}_\phi$, and are illustrated in Fig. 11.2. They are mutually orthogonal. Then

$$(\boldsymbol{e}_s \quad \boldsymbol{e}_\theta \quad \boldsymbol{e}_\phi)^{\mathrm{T}} = \mathbf{R}'(\boldsymbol{e}_r \quad \boldsymbol{e}_\phi \quad \boldsymbol{e}_z)^{\mathrm{T}} = \mathbf{R}''(\boldsymbol{e}_1 \quad \boldsymbol{e}_2 \quad \boldsymbol{e}_3)^{\mathrm{T}} \tag{11.50}$$

where

$$\mathbf{R}' = \begin{pmatrix} \sin\theta & 0 & \cos\theta \\ \cos\theta & 0 & -\sin\theta \\ 0 & 1 & 0 \end{pmatrix},$$

$$\mathbf{R}'' = \begin{pmatrix} \sin\theta\cos\phi & \sin\theta\sin\phi & \cos\theta \\ \cos\theta\cos\phi & \cos\theta\sin\phi & -\sin\theta \\ -\sin\phi & \cos\phi & 0 \end{pmatrix} \tag{11.51}$$

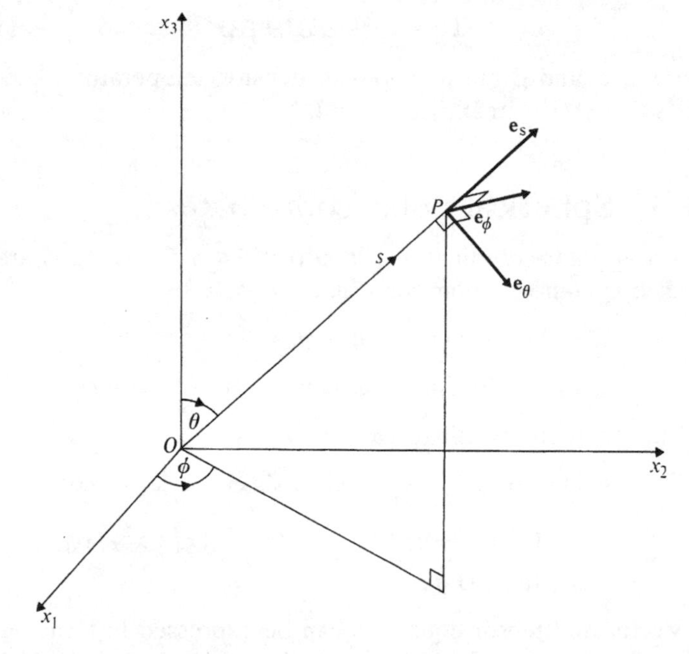

Figure 11.2 Base vectors for spherical polar coordinates

The matrices $\mathbf{R}'$ and $\mathbf{R}''$ are orthogonal matrices and, of course, $\mathbf{R}''=\mathbf{R}'\mathbf{R}$.

If the vector $\boldsymbol{a}$ has components a_s, a_θ, a_ϕ in the system s, θ, ϕ, then

$$\boldsymbol{a}=a_s\boldsymbol{e}_s+a_\theta\boldsymbol{e}_\theta+a_\phi\boldsymbol{e}_\phi$$

and, if $\mathbf{a}^{**}$ denotes the column matrix $(a_s \quad a_\theta \quad a_\phi)^{\mathrm{T}}$. we have

$$\mathbf{a}^{**}=\mathbf{R}'\mathbf{a}^*=\mathbf{R}''\mathbf{a}; \qquad \mathbf{a}=\mathbf{R}''^{\mathrm{T}}\mathbf{a}^{**}, \qquad \mathbf{a}^*=\mathbf{R}'^{\mathrm{T}}\mathbf{a}^{**} \tag{11.52}$$

The second-order tensor $\boldsymbol{A}$ can be expressed in the form

$$\boldsymbol{A}=(\boldsymbol{e}_s \quad \boldsymbol{e}_\theta \quad \boldsymbol{e}_\phi)\otimes\mathbf{A}^{**}(\boldsymbol{e}_s \quad \boldsymbol{e}_\theta \quad \boldsymbol{e}_\phi)^{\mathrm{T}} \tag{11.53}$$

where

$$\mathbf{A}^{**}=\begin{pmatrix} A_{ss} & A_{s\theta} & A_{s\phi} \\ A_{\theta s} & A_{\theta\theta} & A_{\theta\phi} \\ A_{\phi s} & A_{\phi\theta} & A_{\phi\phi} \end{pmatrix}, \tag{11.54}$$

is the matrix of the components, referred to spherical polar coordinates, of the tensor $\boldsymbol{A}$. Then

$$\mathbf{A}^{**}=\mathbf{R}'\mathbf{A}^*\mathbf{R}'^{\mathrm{T}}=\mathbf{R}''\mathbf{A}\mathbf{R}''^{\mathrm{T}}; \qquad \mathbf{A}=\mathbf{R}''^{\mathrm{T}}\mathbf{A}^{**}\mathbf{R}'', \qquad \mathbf{A}^*=\mathbf{R}'^{\mathrm{T}}\mathbf{A}^{**}\mathbf{R}' \tag{11.55}$$

The principal values of $\boldsymbol{A}$ are the roots of $\det(\mathbf{A}^{**}-A\mathbf{I})=0$, and the invariants I_1, I_2 and I_3 of $\boldsymbol{A}$ may be written as

$$I_1=\operatorname{tr}\mathbf{A}^{**}, \qquad I_2=\tfrac{1}{2}\{(\operatorname{tr}\mathbf{A}^{**})^2-\operatorname{tr}\mathbf{A}^{**2}\}, \qquad I_3=\det\mathbf{A}^{**} \tag{11.56}$$

Referred to spherical polar coordinates, the gradient of the scalar $\psi(s, \theta, \phi)$ and the divergence of the vector $\boldsymbol{a}(s, \theta, \phi)$ are

$$\operatorname{grad}\psi=\boldsymbol{\nabla}\psi=\frac{\partial\psi}{\partial s}\boldsymbol{e}_s+\frac{1}{s}\frac{\partial\psi}{\partial\theta}\boldsymbol{e}_\theta+\frac{1}{s\sin\theta}\frac{\partial\psi}{\partial\phi}\boldsymbol{e}_\phi \tag{11.57}$$

$$\operatorname{div}\boldsymbol{a}=\boldsymbol{\nabla}\cdot\boldsymbol{a}=\frac{1}{s^2}\frac{\partial}{\partial s}(s^2a_s)+\frac{1}{s\sin\theta}\frac{\partial}{\partial\theta}(a_\theta\sin\theta)+\frac{1}{s\sin\theta}\frac{\partial a_\phi}{\partial\phi} \tag{11.58}$$

The material derivative of ψ is given by (4.18) as

$$\dot{\psi}=\partial\psi/\partial t+\boldsymbol{v}\cdot\operatorname{grad}\psi$$

and the acceleration vector $\boldsymbol{f}$ is given in terms of the velocity vector $\boldsymbol{v}$ as

$$\boldsymbol{f}=\partial\boldsymbol{v}/\partial t+\left(v_s\frac{\partial}{\partial s}+\frac{v_\theta}{s}\frac{\partial}{\partial\theta}+\frac{v_\phi}{s\sin\theta}\frac{\partial}{\partial\phi}\right)(v_s\boldsymbol{e}_s+v_\theta\boldsymbol{e}_\theta+v_\phi\boldsymbol{e}_\phi)$$

Let the matrix of components, referred to base vectors $\boldsymbol{e}_s$, $\boldsymbol{e}_\theta$, $\boldsymbol{e}_\phi$, of the stress tensor $\boldsymbol{T}$ be $\mathbf{T}^{**}$, where

$$\mathbf{T}^{**}=\begin{pmatrix} T_{ss} & T_{s\theta} & T_{s\phi} \\ T_{\theta s} & T_{\theta\theta} & T_{\theta\phi} \\ T_{\phi s} & T_{\phi\theta} & T_{\phi\phi} \end{pmatrix} \tag{11.59}$$

Then, from (11.55),

$$\mathbf{T}^{**}=\mathbf{R}'\mathbf{T}^{*}\mathbf{R}'^{\mathrm{T}}=\mathbf{R}''\mathbf{T}\mathbf{R}''^{\mathrm{T}}, \qquad \mathbf{T}=\mathbf{R}''^{\mathrm{T}}\mathbf{T}^{**}\mathbf{R}'', \qquad \mathbf{T}^{*}=\mathbf{R}'^{\mathrm{T}}\mathbf{T}^{**}\mathbf{R}' \tag{11.60}$$

and the stress invariants J_1, J_2 and J_3 can be written as

$$J_1=\operatorname{tr}\mathbf{T}^{**}, \qquad J_2=\tfrac{1}{2}\{\operatorname{tr}\mathbf{T}^{**2}-(\operatorname{tr}\mathbf{T}^{**})^2\}, \qquad J_3=\det\mathbf{T}^{**} \tag{11.61}$$

Now consider a finite deformation in which a typical particle which initially has spherical polar coordinates S, Θ, Φ moves to the position with spherical polar coordinates s, θ, ϕ. The motion can be described by equations of the form

$$s=s(S,\Theta,\Phi), \qquad \theta=\theta(S,\Theta,\Phi), \qquad \phi=\phi(S,\Theta,\Phi)$$

In addition to the matrices $\mathbf{R}'$ and $\mathbf{R}''$ defined by (11.51), we introduce orthogonal matrices $\mathbf{R}_0'$ and $\mathbf{R}_0''$, where

$$\mathbf{R}_0'=\begin{pmatrix} \sin\Theta & 0 & \cos\Theta \\ \cos\Theta & 0 & -\sin\Theta \\ 0 & 1 & 0 \end{pmatrix}$$

$$\mathbf{R}_0''=\begin{pmatrix} \sin\Theta\cos\Phi & \sin\Theta\sin\Phi & \cos\Theta \\ \cos\Theta\cos\Phi & \cos\Theta\sin\Phi & -\sin\Theta \\ -\sin\Phi & \cos\Phi & 0 \end{pmatrix} \tag{11.62}$$

Then, after some manipulation, we obtain

$$\mathbf{F}^{**}=\mathbf{R}'\mathbf{F}^{*}\mathbf{R}_0'^{\mathrm{T}}=\mathbf{R}''\mathbf{F}\mathbf{R}_0''^{\mathrm{T}}=\begin{pmatrix} \dfrac{\partial s}{\partial S} & \dfrac{1}{S}\dfrac{\partial s}{\partial\Theta} & \dfrac{1}{S\sin\Theta}\dfrac{\partial s}{\partial\Phi} \\ \dfrac{s\,\partial\theta}{\partial S} & \dfrac{s}{S}\dfrac{\partial\theta}{\partial\Theta} & \dfrac{s}{S\sin\Theta}\dfrac{\partial\theta}{\partial\Phi} \\ s\sin\theta\dfrac{\partial\phi}{\partial S} & \dfrac{s\sin\theta}{S}\dfrac{\partial\phi}{\partial\Theta} & \dfrac{s\sin\theta}{S\sin\Theta}\dfrac{\partial\phi}{\partial\Phi} \end{pmatrix} \tag{11.63}$$

Then the matrices of the components, referred to spherical polar coordinates, of $\boldsymbol{B}$ and $\boldsymbol{C}$ are

$$\mathbf{B}^{**} = \mathbf{R}'\mathbf{B}^{*}\mathbf{R}'^{\mathrm{T}} = \mathbf{R}''\mathbf{B}\mathbf{R}''^{\mathrm{T}} = \mathbf{F}^{**}\mathbf{F}^{**\mathrm{T}} \qquad (11.64)$$

$$\mathbf{C}^{**} = \mathbf{R}_0'\mathbf{C}^{*}\mathbf{R}_0'^{\mathrm{T}} = \mathbf{R}_0''\mathbf{C}\mathbf{R}_0''^{\mathrm{T}} = \mathbf{F}^{**\mathrm{T}}\mathbf{F}^{**} \qquad (11.65)$$

For a small displacement $\mathbf{u} = u_s\mathbf{e}_s + u_\theta\mathbf{e}_\theta + u_\phi\mathbf{e}_\phi$, we have

$$u_s = u_r \sin\theta + u_z \cos\theta, \; u_\theta = u_r\cos\theta - u_z \sin\theta \qquad (11.66)$$

Hence, for small displacements $\tilde{\mathbf{F}} - \mathbf{I} = \mathbf{R}'\hat{\mathbf{F}}\mathbf{R}'^{\mathrm{T}} - \mathbf{I} \simeq$

$$\begin{pmatrix} \dfrac{\partial u_s}{\partial s} & \dfrac{1}{s}\left(\dfrac{\partial u_s}{\partial \theta} - u_\theta\right) & \dfrac{1}{s}\left(\dfrac{1}{\sin\theta}\dfrac{\partial u_s}{\partial \phi} - u_\phi\right) \\ \dfrac{\partial u_\theta}{\partial s} & \dfrac{1}{s}\left(\dfrac{\partial u_\theta}{\partial \theta} + u_s\right) & \dfrac{1}{s}\left(\dfrac{1}{\sin\theta}\dfrac{\partial u_\theta}{\partial \phi} - u_\phi\cot\theta\right) \\ \dfrac{\partial u_\phi}{\partial s} & \dfrac{1\partial u_\phi}{s\partial s} & \dfrac{1}{s}\left(\dfrac{1}{\sin\theta}\dfrac{\partial u_\phi}{\partial \phi} + u_s + u_\theta\cot\theta\right) \end{pmatrix} \qquad (11.67)$$

Then the matrix $\mathbf{E}^{**}$ of infinitesimal strain components, and the matrix Ω^{**} of infinitesimal rotation components, referred to spherical polar coordinates, are given by

$$\mathbf{E}^{**} = \tfrac{1}{2}(\tilde{\mathbf{F}} + \tilde{\mathbf{F}}^{\mathrm{T}}) - \mathbf{I}, \qquad \Omega^{**} = \tfrac{1}{2}(\tilde{\mathbf{F}} - \tilde{\mathbf{F}}^{\mathrm{T}}) \qquad (11.68)$$

Similarly, the matrix $\mathbf{L}^{**}$ of the components, referred to s, θ, ϕ coordinates, of the velocity gradient tensor $\boldsymbol{L}$ is obtained from (11.67) by replacing $\mathbf{F}^{**} - \mathbf{I}$ by $\mathbf{L}^{**}$, u_s, u_θ and u_ϕ by v_s, v_θ and v_ϕ respectively, and S, Θ and Φ by s, θ and ϕ respectively. The expression is exact. The matrices $\mathbf{D}^{**}$ and $\mathbf{W}^{**}$ of the components, referred to s, θ, ϕ coordinates, of the rate-of-deformation tensor $\boldsymbol{D}$ and the vorticity tensor $\boldsymbol{W}$ are then given by

$$\mathbf{D}^{**} = \tfrac{1}{2}(\mathbf{L}^{**} + \mathbf{L}^{**\mathrm{T}}), \qquad \mathbf{W}^{**} = \tfrac{1}{2}(\mathbf{L}^{**} - \mathbf{L}^{**\mathrm{T}}) \qquad (11.69)$$

From (11.63), $\det\mathbf{F} = \det\mathbf{F}^{**}$, and so from (7.8)

$$\frac{\rho_0}{\rho} = \frac{\mathrm{d}v}{\mathrm{d}V} = \det\mathbf{F}^{**} \qquad (11.70)$$

By resolving the body force and acceleration into components referred to base vectors $\boldsymbol{e}_s$, $\boldsymbol{e}_\theta$ and $\boldsymbol{e}_\phi$, the equations of motion can be expressed as

$$\begin{aligned}
&\frac{\partial T_{ss}}{\partial s}+\frac{1}{s}\frac{\partial T_{s\theta}}{\partial \theta}+\frac{1}{s\sin\theta}\frac{\partial T_{s\phi}}{\partial \phi}\\
&\quad +s^{-1}(2T_{ss}-T_{\theta\theta}-T_{\phi\phi}+T_{s\theta}\cot\theta)+\rho b_s=\rho f_s\\
&\frac{\partial T_{s\theta}}{\partial s}+\frac{1}{s}\frac{\partial T_{\theta\theta}}{\partial \theta}+\frac{1}{s\sin\theta}\frac{\partial T_{\theta\phi}}{\partial \phi}\\
&\quad + s^{-1}(T_{\theta\theta}\cot\theta-T_{\phi\phi}\cot\theta+3T_{s\theta})+\rho b_\theta=\rho f_\theta\\
&\frac{\partial T_{s\phi}}{\partial s}+\frac{1}{s}\frac{\partial T_{\theta\phi}}{\partial \theta}+\frac{1}{s\sin\theta}\frac{\partial T_{\phi\phi}}{\partial \phi}\\
&\quad +s^{-1}(3T_{s\phi}+2T_{\theta\phi}\cot\theta)+\rho f_\phi=\rho f_\phi
\end{aligned}\tag{11.71}$$

Alternatively, these equations can be derived by considering the forces acting on an elementary region bounded by the surfaces

$$s=s_0,\qquad s=s_0+\delta s,\qquad \theta=\theta_0,\qquad \theta=\theta_0+\delta\theta$$
$$\phi=\phi_0\qquad\text{and}\qquad \phi=\phi_0+\delta\phi$$

By arguments analogous to those which lead to (11.40) and (11.41), the constitutive equations for an isotropic linear elastic solid, and for a Newtonian viscous fluid, can be expressed as

$$\mathbf{T}^{**}=\lambda\mathbf{I}\operatorname{tr}\mathbf{E}^{**}+2\mu\mathbf{E}^{**}\tag{11.72}$$

and

$$\mathbf{T}^{**}=(-p+\lambda\operatorname{tr}\mathbf{D}^{**})\mathbf{I}+2\mu\mathbf{D}^{**}\tag{11.73}$$

respectively, where in (11.72) λ and μ are elastic constants, and in (11.73) p, λ and μ have the same meaning as in (11.41).

The strain invariants I_1, I_2 and I_3 can be expressed as

$$\begin{aligned}
I_1&=\operatorname{tr}\mathbf{C}^{**}=\operatorname{tr}\mathbf{B}^{**}\\
I_2&=\tfrac{1}{2}\{(\operatorname{tr}\mathbf{C}^{**})^2-\operatorname{tr}\mathbf{C}^{**2}\}=\tfrac{1}{2}\{(\operatorname{tr}\mathbf{B}^{**})^2-\operatorname{tr}\mathbf{B}^{**2}\}\\
&=I_3\operatorname{tr}\mathbf{C}^{**-1}=I_3\operatorname{tr}\mathbf{B}^{**-1}\\
I_3&=\det\mathbf{C}^{**}=\det\mathbf{B}^{**}
\end{aligned}\tag{11.74}$$

The constitutive equation for an isotropic elastic solid can be written as

$$\mathbf{T}^{**}=2(I_3)^{-\frac{1}{2}}\{(I_2W_2+I_3W_3)\mathbf{I}+W_1\mathbf{B}^{**}-I_3W_2\mathbf{B}^{**-1}\}\tag{11.75}$$

or, in the case of an incompressible material, as

$$\mathbf{T}^{**}=-p\mathbf{I}+2W_1\mathbf{B}^{**}-2W_2\mathbf{B}^{**-1}\tag{11.76}$$

The constitutive equation for a Reiner–Rivlin fluid can be expressed in the form

$$\mathbf{T}^{**} = -p\mathbf{I} + \alpha\mathbf{D}^{**} + \beta\mathbf{D}^{**2} \tag{11.77}$$

where p, α and β can be expressed as functions of density, temperature, $\operatorname{tr}\mathbf{D}^{**}$, $\frac{1}{2}\{(\operatorname{tr}\mathbf{D}^{**})^2 - \operatorname{tr}\mathbf{D}^{**2}\}$ and $\det \mathbf{D}^{**}$.

11.4 Problems

1. Steady helical flow is defined by the equations

$$r = R, \qquad \phi = \Phi + t\omega(R), \qquad z = Z + t\alpha(R)$$

where ω and α are functions only of R.

(a) Sketch the path followed by typical particle; (b) find the velocity of the particle at (r, ϕ, z) at time t; (c) find the velocity of the particle which was at (R, Φ, Z) at $t = 0$; (d) find the acceleration of the particle at (r, ϕ, z) at time t; (e) find the divergence of the velocity vector; (f) find the components of $\boldsymbol{L}$, $\boldsymbol{D}$ and $\boldsymbol{\Omega}$ referred to (r, ϕ, z) coordinates.

2. If $v_r = v(r, t)$, $v_\phi = 0$, $v_z = 0$, show that the acceleration vector is directed in the r direction, and has magnitude $\partial v/\partial t + v\, \partial v/\partial r$.

3. If, in cylindrical polar coordinates,

$$r = R\{1 + \tfrac{1}{2}(t/t_0)^2\}, \qquad \phi = \Phi, \qquad z = Z$$

find the velocity and acceleration in terms of r, ϕ, z and t.

4. For the deformation defined by

$$r = AR, \qquad \phi = B \log R + C\Phi, \qquad z = \frac{Z}{A^2C}$$

where A, B and C are constants, determine the matrix $\mathbf{B}^*$, and show that the invariants I_1, I_2, I_3 are constants.

5. If $\boldsymbol{A}$ is the unit vector $A_R\boldsymbol{e}_R + A_\Phi\boldsymbol{e}_\Phi + A_Z\boldsymbol{e}_Z$, and $\mathbf{A}^*$ is the matrix $(A_R \quad A_\Phi \quad A_Z)^{\mathrm{T}}$, show that the extension of a line element which has the direction $\boldsymbol{A}$ in the reference configuration is given by $(\lambda^2) = \mathbf{A}^{*\mathrm{T}}\mathbf{C}^*\mathbf{A}^*$. Hence determine the initial directions of all the line elements whose length does not change in the pure torsion deformation

$$r = R, \qquad \phi = \Phi + \psi Z, \qquad z = Z, \qquad \text{where } \psi \text{ is constant.}$$

6. Prove that the eigenvalues of $\mathbf{C}^*$ are the same as those of $\mathbf{C}$, and that if $\mathbf{y}$ is an eigenvector of $\mathbf{C}$, then $\mathbf{R}_0\mathbf{y}$ is an eigenvector of $\mathbf{C}^*$. Hence find the principal stretches for the pure torsion deformation of Problem 5.

7. Prove that if $\mathbf{F}_1 = \mathbf{RF}$, then

$$\mathbf{F}_1 = \begin{pmatrix} \dfrac{\partial r}{\partial X_1} & \dfrac{\partial r}{\partial X_2} & \dfrac{\partial r}{\partial X_3} \\ r\dfrac{\partial \phi}{\partial X_1} & r\dfrac{\partial \phi}{\partial X_2} & r\dfrac{\partial \phi}{\partial X_3} \\ \dfrac{\partial z}{\partial X_1} & \dfrac{\partial z}{\partial X_2} & \dfrac{\partial z}{\partial X_3} \end{pmatrix}$$

and that $\mathbf{B}^* = \mathbf{F}_1\mathbf{F}_1^{\mathrm{T}}$, $\mathbf{C}^* = \mathbf{F}_1^{\mathrm{T}}\mathbf{F}_1$.

8. Prove that the stress resulting in a compressible isotropic elastic solid from the pure torsion deformation of Problem 5 will not, in general, satisfy the equations of equilibrium.

9. A circular cylinder of isotropic incompressible material undergoes the extension and torsion deformation

$$z = \lambda Z, \qquad r = \lambda^{-\frac{1}{2}}R, \qquad \phi = \Phi + \psi Z$$

where λ and ψ are constants. Find the stress component $T_{\phi z}$ and hence determine the end couple required to maintain the deformation if $W = C_1(I_1 - 3) + C_2(I_2 - 3)$, where C_1 and C_2 are constants.

10. The matrix $\mathbf{F}_2$ is defined as $\mathbf{F}_2 = \mathbf{F}\mathbf{R}_0^{\mathrm{T}}$. Prove that $\mathbf{B} = \mathbf{F}_2\mathbf{F}_2^{\mathrm{T}}$ $\mathbf{C} = \mathbf{F}_2^{\mathrm{T}}\mathbf{F}_2$, and that

$$\mathbf{F}_2 = \begin{pmatrix} \dfrac{\partial x_1}{\partial R} & \dfrac{1}{R}\dfrac{\partial x_1}{\partial \Phi} & \dfrac{\partial x_1}{\partial Z} \\ \dfrac{\partial x_2}{\partial R} & \dfrac{1}{R}\dfrac{\partial x_2}{\partial \Phi} & \dfrac{\partial x_2}{\partial Z} \\ \dfrac{\partial x_3}{\partial R} & \dfrac{1}{R}\dfrac{\partial x_3}{\partial \Phi} & \dfrac{\partial x_3}{\partial Z} \end{pmatrix}$$

An isotropic incompressible elastic body is initially bounded by the surfaces $R = A$, $R = \sqrt{2}A$, $\Phi = \pm\alpha$, $Z = \pm B$, where A, B and α are constants. It undergoes the deformation

$$x_1 = \tfrac{1}{2}R^2/A, \qquad x_2 = A\Phi, \qquad x_3 = Z$$

Sketch the body in its reference and deformed configurations.

Show that the deformation is possible in an incompressible material, and determine the stress in the deformed body.

11. The behaviour of an incompressible non-Newtonian fluid is governed by the constitutive equation

$$\boldsymbol{T} = -p\boldsymbol{I} + 2\mu(1 - 2\varepsilon \operatorname{tr} \boldsymbol{D}^2)\boldsymbol{D} + 4\beta\boldsymbol{D}^2$$

where μ, ε and β are constants with $\varepsilon \ll 1$. Determine the stress components in cylindrical polar coordinates when the fluid is undergoing the flow

$$v_r = 0 \qquad v_\phi = 0, \qquad v_z = w(r)$$

Verify that this is compatible with the incompressibility condition and show that in order to satisfy the equations of motion, $w(r)$ is given by

$$\frac{dw}{dr}\left[1 - \varepsilon\left(\frac{dw}{dr}\right)^2\right] = \frac{c}{r} - \frac{kr}{2\mu}$$

where c is an arbitrary constant and $k = -\partial p/\partial z$. By writing

$$w(r) = w_0(r) + \varepsilon w_1(r) + \varepsilon^2 w_2(r) + \ldots$$

obtain an expression for $w(r)$, correct to terms of order ε, which gives the velocity distribution for axial flow along a circular pipe of radius a, under a constant pressure gradient k.

12. The relations

$$s^3 - a^3 = -(S^3 - A^3), \qquad \theta = \pi - \Theta, \qquad \phi = \Phi$$

where A and a are constants, describe the eversion (turning inside-out) of a sphere. Find $\mathbf{F}^{**}$ and $\mathbf{B}^{**}$ for this deformation. Hence determine the stress in an incompressible isotropic elastic solid with strain-energy function $W = C(I_1 - 3)$, where C is constant.

Appendix

Representation theorem for an isotropic tensor function of a tensor

Suppose that $\boldsymbol{T}$ and $\boldsymbol{D}$ are second-order tensors, such that the components of $\boldsymbol{T}$ are functions of the components of $\boldsymbol{D}$, thus

$$\boldsymbol{T} = \boldsymbol{T}(\boldsymbol{D})$$

Then if

$$\boldsymbol{T}(\boldsymbol{M} \cdot \boldsymbol{D} \cdot \boldsymbol{M}^{\mathrm{T}}) = \boldsymbol{M} \cdot \boldsymbol{T}(\boldsymbol{D}) \cdot \boldsymbol{M}^{\mathrm{T}} \tag{A.1}$$

for all orthogonal tensors $\boldsymbol{M}$, we say that $\boldsymbol{T}(\boldsymbol{D})$ is an isotropic tensor function of $\boldsymbol{D}$. We consider the case in which $\boldsymbol{T}$ and $\boldsymbol{D}$ are symmetric tensors, and denote

$$\bar{\boldsymbol{T}} = \boldsymbol{M} \cdot \boldsymbol{T} \cdot \boldsymbol{M}^{\mathrm{T}}, \qquad \bar{\boldsymbol{D}} = \boldsymbol{M} \cdot \boldsymbol{D} \cdot \boldsymbol{M}^{\mathrm{T}} \tag{A.2}$$

Theorem. $\boldsymbol{T}$ is an isotropic tensor function of $\boldsymbol{D}$ if and only if

$$\boldsymbol{T}(\boldsymbol{D}) = \alpha \boldsymbol{I} + \beta \boldsymbol{D} + \gamma \boldsymbol{D}^2 \tag{A.3}$$

where α, β, γ are scalar functions of $\operatorname{tr} \boldsymbol{D}$, $\operatorname{tr} \boldsymbol{D}^2$ and $\operatorname{tr} \boldsymbol{D}^3$.

Proof (a) *Sufficiency.* Since $\boldsymbol{M}$ is orthogonal, $\operatorname{tr} \boldsymbol{D} = \operatorname{tr} \bar{\boldsymbol{D}}$, $\operatorname{tr} \boldsymbol{D}^2 = \operatorname{tr} \bar{\boldsymbol{D}}^2$, and $\operatorname{tr} \boldsymbol{D}^3 = \operatorname{tr} \bar{\boldsymbol{D}}^3$. Hence α, β and γ are unchanged if D_{ij} are replaced by $\bar{D}_{ij}$.

Assume (A.3) holds. Then from (A.2),

$$\begin{aligned}
\boldsymbol{M} \cdot \boldsymbol{T}(\boldsymbol{D}) \cdot \boldsymbol{M}^{\mathrm{T}} &= \boldsymbol{M} \cdot (\alpha \boldsymbol{I} + \beta \boldsymbol{D} + \gamma \boldsymbol{D}^2) \cdot \boldsymbol{M}^{\mathrm{T}} \\
&= \alpha \boldsymbol{I} + \beta \boldsymbol{M} \cdot \boldsymbol{D} \cdot \boldsymbol{M}^{\mathrm{T}} + \gamma \boldsymbol{M} \cdot \boldsymbol{D} \cdot \boldsymbol{M}^{\mathrm{T}} \cdot \boldsymbol{M} \cdot \boldsymbol{D} \cdot \boldsymbol{M}^{\mathrm{T}} \\
&= \alpha \boldsymbol{I} + \beta \bar{\boldsymbol{D}} + \gamma \bar{\boldsymbol{D}}^2 = \boldsymbol{T}(\bar{\boldsymbol{D}})
\end{aligned}$$

(b) *Necessity.* Assume that (A.1) is satisfied, and choose the x_i coordinate system so that the coordinate axes are the principal axes of $\boldsymbol{D}$. Then, in these coordinates,

$$(D_{ij}) = \begin{pmatrix} D_1 & 0 & 0 \\ 0 & D_2 & 0 \\ 0 & 0 & D_3 \end{pmatrix} \tag{A.4}$$

and

$$T_{ij} = T_{ij}(D_1, D_2, D_3) \tag{A.5}$$

Choose

$$\mathbf{M} = (M_{ij}) = \begin{pmatrix} 1 & 0 & 0 \\ 0 & -1 & 0 \\ 0 & 0 & -1 \end{pmatrix}$$

Then $\bar{D}_{ij} = D_{ij}$, and (A.6)

$$(\bar{T}_{ij}) = \begin{pmatrix} T_{11} & -T_{12} & -T_{13} \\ -T_{12} & T_{22} & T_{23} \\ -T_{13} & T_{23} & T_{33} \end{pmatrix} \tag{A.7}$$

However, (A.1) and (A.6) require that $\bar{T}_{ij} = T_{ij}$. Hence $T_{12} = 0$, $T_{13} = 0$. Similarly, by another choice of $\mathbf{M}$, it can be shown that $T_{23} = 0$. Thus if (D_{ij}) is a diagonal matrix, so is (T_{ij}); that is, $\boldsymbol{D}$ and $\boldsymbol{T}$ have the same principal axes. Therefore we can now write

$$T_{11} = T_1 = F(D_1, D_2, D_3), \qquad T_{22} = T_2 = F_2(D_1, D_2, D_3)$$
$$T_{33} = T_3 = F_3(D_1, D_2, D_3) \tag{A.8}$$

Next choose

$$\mathbf{M} = (M_{ij}) = \begin{pmatrix} 0 & 1 & 0 \\ 0 & 0 & 1 \\ 1 & 0 & 0 \end{pmatrix}$$

Then

$$\mathbf{MDM}^{\mathrm{T}} = \begin{pmatrix} D_2 & 0 & 0 \\ 0 & D_3 & 0 \\ 0 & 0 & D_1 \end{pmatrix}, \qquad \mathbf{MTM}^{\mathrm{T}} = \begin{pmatrix} T_2 & 0 & 0 \\ 0 & T_3 & 0 \\ 0 & 0 & T_1 \end{pmatrix}$$

and so (A.1) gives

$$\begin{aligned} F(D_2, D_3, D_1) &= F_2(D_1, D_2, D_3), \\ F_2(D_2, D_3, D_1) &= F_3(D_1, D_2, D_3), \\ F_3(D_2, D_3, D_1) &= F(D_1, D_2, D_3) \end{aligned} \tag{A.9}$$

Hence T_1, T_2 and T_3 can be expressed in terms of the single function $F(D_1, D_2, D_3)$ as

$$T_1 = F(D_1, D_2, D_3), \qquad T_2 = F(D_2, D_3, D_1), \qquad T_3 = F(D_3, D_1, D_2) \tag{A.10}$$

Finally, choose

$$\mathbf{M} = (M_{ij}) = \begin{pmatrix} 0 & 1 & 0 \\ 1 & 0 & 0 \\ 0 & 0 & -1 \end{pmatrix}$$

Then

$$\mathbf{MDM}^{\mathrm{T}} = \begin{pmatrix} D_2 & 0 & 0 \\ 0 & D_1 & 0 \\ 0 & 0 & D_3 \end{pmatrix}, \qquad \mathbf{MTM}^{\mathrm{T}} = \begin{pmatrix} T_2 & 0 & 0 \\ 0 & T_1 & 0 \\ 0 & 0 & T_3 \end{pmatrix}$$

and then (A.1) gives

$$F(D_2, D_1, D_3) = F(D_2, D_3, D_1) \tag{A.11}$$

Now the equations

$$\begin{aligned} \alpha + \beta D_1 + \gamma D_1^2 &= F(D_1, D_2, D_3) \\ \alpha + \beta D_2 + \gamma D_2^2 &= F(D_2, D_3, D_1) \\ \alpha + \beta D_3 + \gamma D_3^2 &= F(D_3, D_1, D_2) \end{aligned} \tag{A.12}$$

have solutions for α, β and γ as functions of D_1, D_2 and D_3. Also, because $F(D_1, D_2, D_3)$ has the symmetry expressed by (A.11), equations (A.12) are unaltered if any pair of D_1, D_2 and D_3 are interchanged. Hence α, β and γ are symmetric functions of D_1, D_2 and D_3. It follows from a theorem in the theory of symmetric functions that α, β and γ can be expressed as functions of

$$\begin{aligned} &D_1 + D_2 + D_3 = \operatorname{tr} \boldsymbol{D}, \qquad D_1^2 + D_2^2 + D_3^2 = \operatorname{tr} \boldsymbol{D}^2 \\ &D_1^3 + D_2^3 + D_3^3 = \operatorname{tr} \boldsymbol{D}^3 \end{aligned} \tag{A.13}$$

Also, from (A.10) and (A.12),

$$\begin{pmatrix} T_1 & 0 & 0 \\ 0 & T_2 & 0 \\ 0 & 0 & T_3 \end{pmatrix} = \alpha \begin{pmatrix} 1 & 0 & 0 \\ 0 & 1 & 0 \\ 0 & 0 & 1 \end{pmatrix} + \beta \begin{pmatrix} D_1 & 0 & 0 \\ 0 & D_2 & 0 \\ 0 & 0 & D_3 \end{pmatrix} + \gamma \begin{pmatrix} D_1^2 & 0 & 0 \\ 0 & D_2^2 & 0 \\ 0 & 0 & D_3^2 \end{pmatrix},$$

which, with (A.13), is equivalent to (A.3).

Answers

Chapter 4

1. (a) $v_1 = v_2 = v_3 = 1 + 2t$, $f_1 = f_2 = f_3 = 2$
   (b) $v_1 = v_2 = v_3 = (1 + t - 2t^2)/(1 - t^3)$, $f_1 = f_2 = f_3 = 2(1 - t)/(1 - t^3)$
   As $t \to 1$, all particles approach the same line $x_1 = x_2 = x_3$

2. $f_1 = -U^2 x_1$, $f_2 = -U^2 x_2$, $f_3 = 0$
   Helices given parametrically by $x_1 = A \cos Ut + B \sin Ut$, $x_2 = A \sin Ut - B \cos Ut$, $x_3 = Vt + C$, where A, B and C are constants

3. $-2U^2 a^4 (x_1^2 + x_2^2)^{-3}(x_1 \boldsymbol{e}_1 + x_2 \boldsymbol{e}_2)$; streamlines $r = r_0 \sin \theta$,
   $V r_0^3(\theta - \frac{1}{2} \sin 2\theta) = 2Ua^2(z - z_0)$ where $x_1 = r \cos \theta$, $x_2 = r \sin \theta$

4. (a) $-(\frac{1}{3}A - \frac{2}{9})e^{-A}$ (b) $\boldsymbol{f} = -2\boldsymbol{e}_1 - 12\boldsymbol{e}_2 + 6\boldsymbol{e}_3$
   (c) $x_1 = 2 \exp(1 - t^{-1})$, $x_2 = -2t^{-2}$, $x_3 = t^{-2}$
   $dx_1 : dx_2 : dx_3 = x_1 x_3 : x_2^2 t : x_2 x_3 t$. Hence $dx_2/dx_3 = x_2/x_3$

5. $x_1 = X_1(1 + t)^A$, $x_2 = X_2(1 + t)^{2A}$, $x_3 = X_3(1 + t)^{3A}$

Chapter 5

1. (a) $3\boldsymbol{e}_1 + 2\boldsymbol{e}_2 + 2\boldsymbol{e}_3$ (b) $(\boldsymbol{e}_1 - 10\boldsymbol{e}_2 + 6\boldsymbol{e}_3)/(14)^{\frac{1}{2}}$
   (c) $(13\boldsymbol{e}_1 + 10\boldsymbol{e}_2 + 8\boldsymbol{e}_3)/(14)^{\frac{1}{2}}$ (d) 0, 3, 6
   (e) direction ratios $2:-1:-2$, $1:-2:2$, $2:2:1$

$$(\bar{T}_{ij}) = \begin{pmatrix} 3 & 0 & 0 \\ 0 & 0 & 0 \\ 0 & 0 & 6 \end{pmatrix}$$

3. Principal components 2, 1, -3. Direction ratios of principal directions $2:0:1$, $0:1:0$, $1:0:-2$

4. (b) $A + Bh^2 = 0$ (c) $-4ah(A + \frac{1}{3}Bh^2)\boldsymbol{e}_2$

5. (c) $-\frac{4}{3}Ch^3 \boldsymbol{e}_2$, $-\frac{8}{3}Calh^3 \boldsymbol{e}_2$ (d) $\frac{8}{3}Calh^3 \boldsymbol{e}_2$

6. (b) $\frac{1}{2}W\pi mL^{-1}\boldsymbol{e}_1 \sin(\frac{1}{2}\pi x_1/L)\cosh mh -$
$\frac{1}{4}W\pi^2L^{-2}\boldsymbol{e}_2\cos(\frac{1}{2}\pi x_1/L)\sinh mh$, $\frac{1}{2}W\pi mL^{-1}\boldsymbol{e}_2\cosh mx_2$
(c) $Wm^2 \sinh mh$, $-\frac{1}{4}W\pi^2L^{-2}\sinh mh$, $\boldsymbol{e}_1$, $\boldsymbol{e}_2$, $\boldsymbol{e}_3$; $\pm\frac{1}{2}W\pi mL^{-1}$, $(\boldsymbol{e}_1\pm\boldsymbol{e}_2)/2^{\frac{1}{2}}$, $\boldsymbol{e}_3$

7. (c) $-\alpha x_2\boldsymbol{e}_1+\alpha x_1\boldsymbol{e}_2+(\beta+\gamma x_1+\delta x_2)\boldsymbol{e}_3$
(d) 0, $\frac{1}{2}\{\beta\pm(\beta^2+4\alpha^2x_1^2+4\alpha^2x_2^2)^{\frac{1}{2}}\}$. Principal stress direction for intermediate principal stress direction is the radial direction

8. (b) direction ratios $\partial\psi/\partial x_1 : \partial\psi/\partial x_2 : 0$ (i.e. the normals to the surfaces $\psi=$ constant)

Chapter 6

2. **(a) direction ratios $7\sqrt{2}:\sqrt{2}-1:\sqrt{2}+1$ (b) $\sqrt{(13/6)}$**

3. $$(F_{iR})=\begin{pmatrix} a_1 & \alpha a_1 & 0 \\ 0 & a_2 & 0 \\ 0 & 0 & a_3 \end{pmatrix}, \qquad (C_{RS})=\begin{pmatrix} a_1^2 & \alpha a_1^2 & 0 \\ \alpha a_1^2 & \alpha^2a_1^2+a_2^2 & 0 \\ 0 & 0 & a_3^2 \end{pmatrix}$$

$$(B_{ij})=\begin{pmatrix} (1+\alpha^2)a_1^2 & \alpha a_1a_2 & 0 \\ \alpha a_1a_2 & a_2^2 & 0 \\ 0 & 0 & a_3^2 \end{pmatrix},$$

$$(F_{iR})^{-1}=\begin{pmatrix} a_1^{-1} & -\alpha a_2^{-1} & 0 \\ 0 & a_2^{-1} & 0 \\ 0 & 0 & a_3^{-1} \end{pmatrix}$$

$$(C_{RS})^{-1}=\begin{pmatrix} a_1^{-2}+\alpha^2a_2^{-2} & -\alpha a_2^{-2} & 0 \\ -\alpha a_2^{-2} & a_2^{-2} & 0 \\ 0 & 0 & a_3^{-2} \end{pmatrix},$$

$$(B_{ij})^{-1}=\begin{pmatrix} a_1^{-2} & -\alpha a_1^{-1}a_2^{-1} & 0 \\ -\alpha a_1^{-1}a_2^{-1} & a_2^{-2}(1+\alpha^2) & 0 \\ 0 & 0 & a_3^{-2} \end{pmatrix}$$

$(\gamma_{RS})=\frac{1}{2}(C_{RS})-\frac{1}{2}(\delta_{RS})$, $(\eta_{ij})=\frac{1}{2}(\delta_{ij})-\frac{1}{2}(B_{ij})^{-1}$, $a_1a_2a_3=1$

Lengths a_1, $(\alpha^2a_1^2+a_2^2)^{\frac{1}{2}}$, a_3; angles $\frac{1}{2}\pi$, $\frac{1}{2}\pi$, $\cos^{-1}\{\alpha a_1/(\alpha^2a_1^2+a_2^2)^{\frac{1}{2}}\}$

4. $\lambda\mu^2=1$, $(\lambda^2+\mu^2\psi^2A^2)^{\frac{1}{2}}$, $\lambda^{-1}(1+A^2\psi^2)^{\frac{1}{2}}$

5. Stretches μ, 1, μ^{-1}. Direction ratios $1:0:0$, $0:0:1$, $\mu\tan\gamma : \mu^{-1}-\mu : 0$

7. $$(E_{ij})=\frac{1}{2}\begin{pmatrix}0 & 0 & -\psi x_2\\ 0 & 0 & \psi x_1\\ -\psi x_2 & \psi x_1 & 0\end{pmatrix},\quad (\Omega_{ij})=\begin{pmatrix}0 & -\psi x_3 & -\frac{1}{2}\psi x_2\\ \psi x_3 & 0 & \frac{1}{2}\psi x_1\\ \frac{1}{2}\psi x_2 & -\frac{1}{2}\psi x_1 & 0\end{pmatrix}$$

Principal components 0, $\pm\frac{1}{2}\psi(x_1^2+x_2^2)^{\frac{1}{2}}$; direction ratios of principal axes $x_1:x_2:0$, $-x_2:x_1:(x_1^2+x_2^2)^{\frac{1}{2}}$, $x_2:-x_1:(x_1^2+x_2^2)^{\frac{1}{2}}$

8. $$(F_{iR}-\delta_{iR})=(E_{iR})=\begin{pmatrix}A-\dfrac{B(X_1^2-X_2^2)}{(X_1^2+X_2^2)^2} & -\dfrac{2BX_1X_2}{(X_1^2+X_2^2)^2} & 0\\ -\dfrac{2BX_1X_2}{(X_1^2+X_2^2)^2} & A+\dfrac{B(X_1^2-X_2^2)}{(X_1^2+X_2^2)^2} & 0\\ 0 & 0 & C\end{pmatrix}$$

$(\Omega_{iR})=\mathbf{0}$; $A\pm B(X_1^2+X_2^2)^{-1}$, C; direction ratios of principal axes $X_2:-X_1:0$, $X_1:X_2:0$, $0:0:1$

9. $$(L_{ij})=(W_{ij})=\begin{pmatrix}0 & -U & 0\\ U & 0 & 0\\ 0 & 0 & 0\end{pmatrix},\quad (D_{ij})=\mathbf{0}$$

$$(L_{ij})=(D_{ij})=\frac{2Ua^2}{(x_1^2+x_2^2)^3}\begin{pmatrix}-x_1(x_1^2-3x_2^2) & -x_2(3x_1^2-x_2^2) & 0\\ -x_2(3x_1^2-x_2^2) & x_1(x_1^2-3x_2^2) & 0\\ 0 & 0 & 0\end{pmatrix}$$

$(W_{ij})=\mathbf{0}$

Chapter 8

2. $2W=\lambda E_{ii}E_{kk}+2\mu E_{ik}E_{ik}+2\alpha E_{ii}E_{33}+4\beta E_{i3}E_{i3}+\gamma E_{33}^2$ (several equivalent alternative forms exist)

8. $S_{ij}=2\mu_0(E_{ij}+t_0D_{ij})$

9. $T_{ij}=-p\delta_{ij}+2\mu_1\int_{-\infty}^{t}\exp\{-(t-\tau)/t_1\}\dot{E}_{ij}(\tau)\,\mathrm{d}\tau$

Chapter 9

1. (a) $\dfrac{7\sqrt{2}}{12}:\dfrac{1}{12}+\sqrt{2}:-\dfrac{1}{12}+\sqrt{2}$ (b) $\dfrac{4}{13}$ (c) $2, \dfrac{3}{2}, \dfrac{1}{2}$
(d) $1:0:0$, $0:1:0$, $0:0:1$; $\frac{1}{2}\sqrt{2}:-\frac{1}{2}:\frac{1}{2}$, $\frac{1}{2}\sqrt{2}:\frac{1}{2}:-\frac{1}{2}$, $0:\frac{1}{2}\sqrt{2}:\frac{1}{2}\sqrt{2}$

2. $$(C_{RS})=\frac{a^2}{R^2}\begin{pmatrix}X_1^2+4X_2^2 & -3X_1X_2 & 0\\ -3X_1X_2 & 4X_1^2+X_2^2 & 0\\ 0 & 0 & b^2R^2/a^2\end{pmatrix},\quad R^2=X_1^2+X_2^2.$$

Principal stretches a, $2a$, b. Direction ratios of principal axes $X_1:X_2:0$, $\quad -X_2:X_1:0$, $\quad 0:0:1$

3. $(B_{ij})^{-1}$

$$=\begin{pmatrix} A^2x_1^2+A^{-2}r^{-4}\lambda^2x_2^2 & A^2x_1x_2-A^{-2}r^{-4}\lambda^2x_1x_2 & 0 \\ A^2x_1x_2-A^{-2}r^{-4}\lambda^2x_1x_2 & A^2x_2^2+A^{-2}r^{-4}\lambda^2x_1^2 & 0 \\ 0 & 0 & \lambda^{-2} \end{pmatrix}$$

4. $$(C_{RS})=\begin{pmatrix} 2\alpha^2 & 0 & 0 \\ 0 & 2\beta^2 & 0 \\ 0 & 0 & \mu^2 \end{pmatrix}, \qquad (R_{iR})=\frac{1}{\sqrt{2}}\begin{pmatrix} 1 & 1 & 0 \\ -1 & 1 & 0 \\ 0 & 0 & \sqrt{2} \end{pmatrix}$$

$$(U_{RS})=\begin{pmatrix} \sqrt{2}\alpha & 0 & 0 \\ 0 & \sqrt{2}\beta & 0 \\ 0 & 0 & \mu \end{pmatrix}$$; principal stretches $\sqrt{2}\alpha$, $\sqrt{2}\beta$, μ

5. $x_1(\tau)=x_1-\alpha(t-\tau)x_2+\alpha\beta(t-\tau)^2x_3$
$x_2(\tau)=x_2-2\beta(t-\tau)x_3$, $x_3(\tau)=x_3$

$$(C_{ij}(\tau))=\begin{pmatrix} 1 & -\alpha s & \alpha\beta s^2 \\ -\alpha s & 1+\alpha^2s^2 & -2\beta s-\alpha^2\beta s^3 \\ \alpha\beta s^2 & -2\beta s-\alpha^2\beta s^3 & 1+4\beta^2s^2+\alpha^2\beta^2s^4 \end{pmatrix}$$

$$A_{ij}^{(0)}=\delta_{ij}, \qquad (A_{ij}^{(1)})=\begin{pmatrix} 0 & \alpha & 0 \\ \alpha & 0 & 2\beta \\ 0 & 2\beta & 0 \end{pmatrix},$$

$$(A_{ij}^{(2)})=2\begin{pmatrix} 0 & 0 & \alpha\beta \\ 0 & \alpha^2 & 0 \\ \alpha\beta & 0 & 4\beta^2 \end{pmatrix}, \qquad (A_{ij}^{(3)})=6\begin{pmatrix} 0 & 0 & 0 \\ 0 & 0 & \alpha^2\beta \\ 0 & \alpha^2\beta & 0 \end{pmatrix},$$

$$(A_{ij}^{(4)})=24\begin{pmatrix} 0 & 0 & 0 \\ 0 & 0 & 0 \\ 0 & 0 & \alpha^2\beta^2 \end{pmatrix}, \qquad A_{ij}^{(n)}=0 \quad (n\geqslant 5)$$

6. $$(A_{ij}^{(1)})=v'(x_2)\begin{pmatrix} 0 & 1 & 0 \\ 1 & 0 & 0 \\ 0 & 0 & 0 \end{pmatrix}, \qquad (A_{ij}^{(2)})=2v'(x_2)^2\begin{pmatrix} 0 & 0 & 0 \\ 0 & 1 & 0 \\ 0 & 0 & 0 \end{pmatrix}$$

Chapter 10

1. Edge lengths λ, $(\alpha^2+\lambda^{-2})^{\frac{1}{2}}$, 1
$T_{11}=-p+2W_1(\lambda^2+\alpha^2)-2W_2\lambda^{-2}$, $T_{12}=2(W_1+W_2)\alpha\lambda^{-1}$,
$T_{13}=0$
$T_{22}=-p+2W_1\lambda^{-2}-2W_2(\lambda^2+\alpha^2)$, $T_{23}=0$,
$T_{33}=-p+2(W_1-W_2)$;
$\lambda(T_{12}\boldsymbol{e}_1+T_{22}\boldsymbol{e}_2)$; $\boldsymbol{n}=(\boldsymbol{e}_1-\alpha\lambda\boldsymbol{e}_2)(1+\alpha^2\lambda^2)^{-\frac{1}{2}}$;
$\{(T_{11}-\alpha\lambda T_{12})\boldsymbol{e}_1+(T_{12}-\alpha\lambda T_{22})\boldsymbol{e}_2\}(1+\alpha^2\lambda^2)^{-\frac{1}{2}}$

2. Edge lengths λ, λ^{-1}, 1
$T_{11}=-p+2\lambda^2C_1-2\lambda^{-2}C_2$, $T_{22}=-p+2\lambda^{-2}C_1-2\lambda^2C_2$,
$T_{33}=-p+2C_1-2C_2$, $T_{23}=T_{31}=T_{12}=0$;
$F_1=\lambda^{-1}T_{11}$, $F_2=\lambda T_{22}$, $F_3=T_{33}$
$\lambda=1$, $2C_2\lambda=C_1-C_2\pm\{(C_1-C_2)^2-4C_2^2\}^{\frac{1}{2}}$

4. $T_{ij}=(\rho/\rho_0)(\partial x_i/\partial X_R)(\partial x_j/\partial X_S)\{4\alpha C_{PP}\delta_{RS}+4\beta C_{RS}+4\gamma C_{11}\delta_{1R}\delta_{1S}$
$\quad+\delta(C_{12}\delta_{1R}\delta_{2S}+C_{12}\delta_{1S}\delta_{2R}+C_{13}\delta_{1R}\delta_{3S}+C_{13}\delta_{1S}\delta_{3R})\}$
$T_{11}=4\lambda(3\alpha+\beta+\gamma)$, $T_{22}=T_{33}=4\lambda(3\alpha+\beta)$,
$T_{23}=T_{31}=T_{12}=0$

5. $\boldsymbol{\chi}=\alpha\boldsymbol{I}+\beta\boldsymbol{B}+\gamma\boldsymbol{B}^2$, where α, β, γ are functions of tr $\boldsymbol{B}$, tr $\boldsymbol{B}^2$ and tr $\boldsymbol{B}^3$

8. $\dfrac{d}{dx_2}\{\alpha(v'^2)v'\}$, $\pm\alpha(v'^2)v'$, where α is a function of v'^2

9. $T_{11}=-p+\frac{1}{4}\beta x_2^2\omega'^2$, $T_{22}=-p+\frac{1}{4}\beta x_1^2\omega'^2$, $T_{33}=-p+\frac{1}{4}\beta(x_1^2+x_2^2)\omega'^2$,
$T_{23}=\frac{1}{2}\mu\{1+\frac{1}{2}\alpha\omega'^2(x_1^2+x_2^2)\}x_1\omega'$,
$T_{13}=-\frac{1}{2}\mu\{1+\frac{1}{2}\alpha\omega'^2(x_1^2+x_2^2)\}x_2\omega'$,
$T_{12}=-\frac{1}{4}\beta x_1x_2\omega'^2$; $A=\Omega/h$, $B=0$

10. $k(U/h)^n$; $k(U/h)^{(n-1)}$

12. $$(C_{ij}(\tau))=\begin{pmatrix}1+\gamma^2s^2r^{-4}x_2^2 & -\gamma^2s^2r^{-4}x_1x_2 & \gamma sr^{-2}x_2\\ -\gamma^2s^2r^{-4}x_1x_2 & 1+\gamma^2s^2r^{-4}x_1^2 & -\gamma sr^{-2}x_1\\ \gamma sr^{-2}x_2 & -\gamma sr^{-2}x_1 & 1\end{pmatrix}$$

$$(A_{ij}^{(1)})=\frac{\gamma}{r^2}\begin{pmatrix}0 & 0 & -x_2\\ 0 & 0 & x_1\\ -x_2 & x_1 & 0\end{pmatrix},\quad (A_{ij}^{(2)})=\frac{2\gamma^2}{r^4}\begin{pmatrix}x_2^2 & -x_1x_2 & 0\\ -x_1x_2 & x_1^2 & 0\\ 0 & 0 & 0\end{pmatrix}$$

13. $T_{11}=T_{22}=T_{33}=-p$, $T_{23}=T_{31}=0$

$$T_{12}=\frac{\mu\omega}{k(k^2+\omega^2)}\{f'(x_2)(-\omega\cos\omega t+k\sin\omega t)-g'(x_2)(\omega\sin\omega t+k\cos\omega t)\}$$

Chapter 11

1. (b) $r\omega(r)\boldsymbol{e}_\phi + \alpha(r)\boldsymbol{e}_z$ (c) $R\omega(R)\boldsymbol{e}_\Phi + \alpha(R)\boldsymbol{e}_Z$ (d) $-r\omega^2(r)\boldsymbol{e}_r$
(e) 0

$$\text{(f) } \mathbf{L}^* = \begin{pmatrix} 0 & 0 & 0 \\ r\omega' & 0 & 0 \\ \alpha' & 0 & 0 \end{pmatrix}, \quad 2\mathbf{D}^* = \begin{pmatrix} 0 & r\omega' & \alpha' \\ r\omega' & 0 & 0 \\ \alpha' & 0 & 0 \end{pmatrix},$$

$$2\mathbf{\Omega}^* = \begin{pmatrix} 0 & -r\omega' & -\alpha' \\ r\omega'' & 0 & 0 \\ \alpha' & 0 & 0 \end{pmatrix}$$

3. $\boldsymbol{v} = rt\boldsymbol{e}_r/(t_0^2 + \frac{1}{2}t^2)$, $\boldsymbol{f} = r\boldsymbol{e}_r/(t_0^2 + \frac{1}{2}t^2)$

4. $$\mathbf{B}^* = \begin{pmatrix} A^2 & A^2B & 0 \\ A^2B & A^2(B^2+C^2) & 0 \\ 0 & 0 & A^{-4}C^{-2} \end{pmatrix}$$

5. $A_z = 0$ or $A_\Phi/A_Z = -\frac{1}{2}R\psi$

6. $1, \{1 + \frac{1}{2}r^2\psi^2 \pm r\psi(1 + \frac{1}{4}r^2\psi^2)^{\frac{1}{2}}\}^{\frac{1}{2}}$

9. $T_{\phi z} = 2(\lambda C_1 + C_2)r\psi$, $\pi a^4\psi(\lambda C_1 + C_2)$, where a is the final radius

10. $T_{11} = -p + 4W_1x_1/A - W_2A/x_1$, $T_{33} = -p + 2(W_1 - W_2)$,
$T_{22} = -p + W_1A/x_1 - 4W_2x_1/A$, $T_{23} = T_{31} = T_{12} = 0$;
$I_1 = I_2 = 2x_1/A + A/2x_1$

11. $T_{rr} = T_{zz} = -p + \beta w'^2$, $T_{\phi\phi} = -p$, $T_{r\phi} = T_{\phi z} = 0$,
$T_{rz} = \mu(1 - \varepsilon w'^2)w'$
$w = -k(r^2 - a^2)/4\mu - \varepsilon k^3(r^4 - a^4)/32\mu^3$

12. $$\mathbf{F}^{**} = \begin{pmatrix} -S^2/s^2 & 0 & 0 \\ 0 & -s/S & 0 \\ 0 & 0 & s/S \end{pmatrix}, \quad \mathbf{B}^{**} = \begin{pmatrix} S^4/s^4 & 0 & 0 \\ 0 & s^2/S^2 & 0 \\ 0 & 0 & s^2/S^2 \end{pmatrix}$$

$T_{ss} = -p + 2CS^4/s^4$, $T_{\theta\phi} = T_{\phi s} = T_{s\theta} = 0$,
$T_{\theta\theta} = T_{\phi\phi} = -p + 2Cs^2/S^2$

Further reading

Chadwick, P. *Continuum Mechanics, Concise Theory and Problems*, George Allen and Unwin, 1976.

Eringen, A. C. *Mechanics of Continua*, Wiley, 1967.

Hunter, S. C. *Mechanics of Continuous Media*, Ellis Horwood, 1976.

Malvern, L. E. *Introduction to the Mechanics of a Continuous Medium*, Prentice Hall, 1969.

Rivlin, R. S. *Non-linear Continuum Theories in Mechanics and Physics and Their Applications*, Edizioni Cremonese, 1970.

Truesdell, C. S. *The Elements of Continuum Mechanics*, Springer, 1966.

In addition to the above texts, which are concerned with continuum mechanics in general, there are many books which deal with particular branches of continuum mechanics, such as elasticity, viscous fluid mechanics, viscoelasticity and so on.

Index

Astronomy

BURNHAM'S CELESTIAL HANDBOOK, Robert Burnham, Jr. Thorough guide to the stars beyond our solar system. Exhaustive treatment. Alphabetical by constellation: Andromeda to Cetus in Vol. 1; Chamaeleon to Orion in Vol. 2; and Pavo to Vulpecula in Vol. 3. Hundreds of illustrations. Index in Vol. 3. 2,000pp. 6⅛ x 9¼.
Vol. I: 0-486-23567-X
Vol. II: 0-486-23568-8
Vol. III: 0-486-23673-0

EXPLORING THE MOON THROUGH BINOCULARS AND SMALL TELESCOPES, Ernest H. Cherrington, Jr. Informative, profusely illustrated guide to locating and identifying craters, rills, seas, mountains, other lunar features. Newly revised and updated with special section of new photos. Over 100 photos and diagrams. 240pp. 8¼ x 11. 0-486-24491-1

THE EXTRATERRESTRIAL LIFE DEBATE, 1750–1900, Michael J. Crowe. First detailed, scholarly study in English of the many ideas that developed from 1750 to 1900 regarding the existence of intelligent extraterrestrial life. Examines ideas of Kant, Herschel, Voltaire, Percival Lowell, many other scientists and thinkers. 16 illustrations. 704pp. 5⅜ x 8½. 0-486-40675-X

THEORIES OF THE WORLD FROM ANTIQUITY TO THE COPERNICAN REVOLUTION, Michael J. Crowe. Newly revised edition of an accessible, enlightening book re-creates the change from an earth-centered to a sun-centered conception of the solar system. 242pp. 5⅜ x 8½. 0-486-41444-2

ARISTARCHUS OF SAMOS: The Ancient Copernicus, Sir Thomas Heath. Heath's history of astronomy ranges from Homer and Hesiod to Aristarchus and includes quotes from numerous thinkers, compilers, and scholasticists from Thales and Anaximander through Pythagoras, Plato, Aristotle, and Heraclides. 34 figures. 448pp. 5⅜ x 8½. 0-486-43886-4

A COMPLETE MANUAL OF AMATEUR ASTRONOMY: TOOLS AND TECHNIQUES FOR ASTRONOMICAL OBSERVATIONS, P. Clay Sherrod with Thomas L. Koed. Concise, highly readable book discusses: selecting, setting up and maintaining a telescope; amateur studies of the sun; lunar topography and occultations; observations of Mars, Jupiter, Saturn, the minor planets and the stars; an introduction to photoelectric photometry; more. 1981 ed. 124 figures. 25 halftones. 37 tables. 335pp. 6½ x 9¼. 0-486-42820-8

AMATEUR ASTRONOMER'S HANDBOOK, J. B. Sidgwick. Timeless, comprehensive coverage of telescopes, mirrors, lenses, mountings, telescope drives, micrometers, spectroscopes, more. 189 illustrations. 576pp. 5⅝ x 8¼. (Available in U.S. only.)
0-486-24034-7

STAR LORE: Myths, Legends, and Facts, William Tyler Olcott. Captivating retellings of the origins and histories of ancient star groups include Pegasus, Ursa Major, Pleiades, signs of the zodiac, and other constellations. "Classic."–Sky & Telescope. 58 illustrations. 544pp. 5⅜ x 8½. 0-486-43581-4

Chemistry

THE SCEPTICAL CHYMIST: THE CLASSIC 1661 TEXT, Robert Boyle. Boyle defines the term "element," asserting that all natural phenomena can be explained by the motion and organization of primary particles. 1911 ed. viii+232pp. 5⅜ x 8½. 0-486-42825-7

RADIOACTIVE SUBSTANCES, Marie Curie. Here is the celebrated scientist's doctoral thesis, the prelude to her receipt of the 1903 Nobel Prize. Curie discusses establishing atomic character of radioactivity found in compounds of uranium and thorium; extraction from pitchblende of polonium and radium; isolation of pure radium chloride; determination of atomic weight of radium; plus electric, photographic, luminous, heat, color effects of radioactivity. ii+94pp. 5⅜ x 8½. 0-486-42550-9

CHEMICAL MAGIC, Leonard A. Ford. Second Edition, Revised by E. Winston Grundmeier. Over 100 unusual stunts demonstrating cold fire, dust explosions, much more. Text explains scientific principles and stresses safety precautions. 128pp. 5⅜ x 8½. 0-486-67628-5

MOLECULAR THEORY OF CAPILLARITY, J. S. Rowlinson and B. Widom. History of surface phenomena offers critical and detailed examination and assessment of modern theories, focusing on statistical mechanics and application of results in mean-field approximation to model systems. 1989 edition. 352pp. 5⅜ x 8½. 0-486-42544-4

CHEMICAL AND CATALYTIC REACTION ENGINEERING, James J. Carberry. Designed to offer background for managing chemical reactions, this text examines behavior of chemical reactions and reactors; fluid-fluid and fluid-solid reaction systems; heterogeneous catalysis and catalytic kinetics; more. 1976 edition. 672pp. 6⅛ x 9¼. 0-486-41736-0 $31.95

ELEMENTS OF CHEMISTRY, Antoine Lavoisier. Monumental classic by founder of modern chemistry in remarkable reprint of rare 1790 Kerr translation. A must for every student of chemistry or the history of science. 539pp. 5⅜ x 8½. 0-486-64624-6

MOLECULES AND RADIATION: An Introduction to Modern Molecular Spectroscopy. Second Edition, Jeffrey I. Steinfeld. This unified treatment introduces upper-level undergraduates and graduate students to the concepts and the methods of molecular spectroscopy and applications to quantum electronics, lasers, and related optical phenomena. 1985 edition. 512pp. 5⅜ x 8½. 0-486-44152-0

A SHORT HISTORY OF CHEMISTRY, J. R. Partington. Classic exposition explores origins of chemistry, alchemy, early medical chemistry, nature of atmosphere, theory of valency, laws and structure of atomic theory, much more. 428pp. 5⅜ x 8½. (Available in U.S. only.) 0-486-65977-1

GENERAL CHEMISTRY, Linus Pauling. Revised 3rd edition of classic first-year text by Nobel laureate. Atomic and molecular structure, quantum mechanics, statistical mechanics, thermodynamics correlated with descriptive chemistry. Problems. 992pp. 5⅜ x 8½. 0-486-65622-5

ELECTRON CORRELATION IN MOLECULES, S. Wilson. This text addresses one of theoretical chemistry's central problems. Topics include molecular electronic structure, independent electron models, electron correlation, the linked diagram theorem, and related topics. 1984 edition. 304pp. 5⅜ x 8½. 0-486-45879-2

Engineering

DE RE METALLICA, Georgius Agricola. The famous Hoover translation of greatest treatise on technological chemistry, engineering, geology, mining of early modern times (1556). All 289 original woodcuts. 638pp. $6\frac{3}{4} \times 11$. 0-486-60006-8

FUNDAMENTALS OF ASTRODYNAMICS, Roger Bate et al. Modern approach developed by U.S. Air Force Academy. Designed as a first course. Problems, exercises. Numerous illustrations. 455pp. $5\frac{3}{8} \times 8\frac{1}{2}$. 0-486-60061-0

DYNAMICS OF FLUIDS IN POROUS MEDIA, Jacob Bear. For advanced students of ground water hydrology, soil mechanics and physics, drainage and irrigation engineering and more. 335 illustrations. Exercises, with answers. 784pp. $6\frac{1}{8} \times 9\frac{1}{4}$. 0-486-65675-6

THEORY OF VISCOELASTICITY (SECOND EDITION), Richard M. Christensen. Complete consistent description of the linear theory of the viscoelastic behavior of materials. Problem-solving techniques discussed. 1982 edition. 29 figures. xiv+364pp. $6\frac{1}{8} \times 9\frac{1}{4}$. 0-486-42880-X

MECHANICS, J. P. Den Hartog. A classic introductory text or refresher. Hundreds of applications and design problems illuminate fundamentals of trusses, loaded beams and cables, etc. 334 answered problems. 462pp. $5\frac{3}{8} \times 8\frac{1}{2}$. 0-486-60754-2

MECHANICAL VIBRATIONS, J. P. Den Hartog. Classic textbook offers lucid explanations and illustrative models, applying theories of vibrations to a variety of practical industrial engineering problems. Numerous figures. 233 problems, solutions. Appendix. Index. Preface. 436pp. $5\frac{3}{8} \times 8\frac{1}{2}$. 0-486-64785-4

STRENGTH OF MATERIALS, J. P. Den Hartog. Full, clear treatment of basic material (tension, torsion, bending, etc.) plus advanced material on engineering methods, applications. 350 answered problems. 323pp. $5\frac{3}{8} \times 8\frac{1}{2}$. 0-486-60755-0

A HISTORY OF MECHANICS, René Dugas. Monumental study of mechanical principles from antiquity to quantum mechanics. Contributions of ancient Greeks, Galileo, Leonardo, Kepler, Lagrange, many others. 671pp. $5\frac{3}{8} \times 8\frac{1}{2}$. 0-486-65632-2

STABILITY THEORY AND ITS APPLICATIONS TO STRUCTURAL MECHANICS, Clive L. Dym. Self-contained text focuses on Koiter postbuckling analyses, with mathematical notions of stability of motion. Basing minimum energy principles for static stability upon dynamic concepts of stability of motion, it develops asymptotic buckling and postbuckling analyses from potential energy considerations, with applications to columns, plates, and arches. 1974 ed. 208pp. $5\frac{3}{8} \times 8\frac{1}{2}$. 0-486-42541-X

BASIC ELECTRICITY, U.S. Bureau of Naval Personnel. Originally a training course; best nontechnical coverage. Topics include batteries, circuits, conductors, AC and DC, inductance and capacitance, generators, motors, transformers, amplifiers, etc. Many questions with answers. 349 illustrations. 1969 edition. 448pp. $6\frac{1}{2} \times 9\frac{1}{4}$. 0-486-20973-3

ROCKETS, Robert Goddard. Two of the most significant publications in the history of rocketry and jet propulsion: "A Method of Reaching Extreme Altitudes" (1919) and "Liquid Propellant Rocket Development" (1936). 128pp. 5⅜ x 8½. 0-486-42537-1

STATISTICAL MECHANICS: PRINCIPLES AND APPLICATIONS, Terrell L. Hill. Standard text covers fundamentals of statistical mechanics, applications to fluctuation theory, imperfect gases, distribution functions, more. 448pp. 5⅜ x 8½. 0-486-65390-0

ENGINEERING AND TECHNOLOGY 1650–1750: ILLUSTRATIONS AND TEXTS FROM ORIGINAL SOURCES, Martin Jensen. Highly readable text with more than 200 contemporary drawings and detailed engravings of engineering projects dealing with surveying, leveling, materials, hand tools, lifting equipment, transport and erection, piling, bailing, water supply, hydraulic engineering, and more. Among the specific projects outlined-transporting a 50-ton stone to the Louvre, erecting an obelisk, building timber locks, and dredging canals. 207pp. 8⅜ x 11¼. 0-486-42232-1

THE VARIATIONAL PRINCIPLES OF MECHANICS, Cornelius Lanczos. Graduate level coverage of calculus of variations, equations of motion, relativistic mechanics, more. First inexpensive paperbound edition of classic treatise. Index. Bibliography. 418pp. 5⅜ x 8½. 0-486-65067-7

PROTECTION OF ELECTRONIC CIRCUITS FROM OVERVOLTAGES, Ronald B. Standler. Five-part treatment presents practical rules and strategies for circuits designed to protect electronic systems from damage by transient overvoltages. 1989 ed. xxiv+434pp. 6⅛ x 9¼. 0-486-42552-5

ROTARY WING AERODYNAMICS, W. Z. Stepniewski. Clear, concise text covers aerodynamic phenomena of the rotor and offers guidelines for helicopter performance evaluation. Originally prepared for NASA. 537 figures. 640pp. 6⅛ x 9¼. 0-486-64647-5

INTRODUCTION TO SPACE DYNAMICS, William Tyrrell Thomson. Comprehensive, classic introduction to space-flight engineering for advanced undergraduate and graduate students. Includes vector algebra, kinematics, transformation of coordinates. Bibliography. Index. 352pp. 5⅜ x 8½. 0-486-65113-4

HISTORY OF STRENGTH OF MATERIALS, Stephen P. Timoshenko. Excellent historical survey of the strength of materials with many references to the theories of elasticity and structure. 245 figures. 452pp. 5⅜ x 8½. 0-486-61187-6

ANALYTICAL FRACTURE MECHANICS, David J. Unger. Self-contained text supplements standard fracture mechanics texts by focusing on analytical methods for determining crack-tip stress and strain fields. 336pp. 6⅛ x 9¼. 0-486-41737-9

STATISTICAL MECHANICS OF ELASTICITY, J. H. Weiner. Advanced, self-contained treatment illustrates general principles and elastic behavior of solids. Part 1, based on classical mechanics, studies thermoelastic behavior of crystalline and polymeric solids. Part 2, based on quantum mechanics, focuses on interatomic force laws, behavior of solids, and thermally activated processes. For students of physics and chemistry and for polymer physicists. 1983 ed. 96 figures. 496pp. 5⅜ x 8½. 0-486-42260-7

Mathematics

FUNCTIONAL ANALYSIS (Second Corrected Edition), George Bachman and Lawrence Narici. Excellent treatment of subject geared toward students with background in linear algebra, advanced calculus, physics and engineering. Text covers introduction to inner-product spaces, normed, metric spaces, and topological spaces; complete orthonormal sets, the Hahn-Banach Theorem and its consequences, and many other related subjects. 1966 ed. 544pp. 6⅛ x 9¼. 0-486-40251-7

DIFFERENTIAL MANIFOLDS, Antoni A. Kosinski. Introductory text for advanced undergraduates and graduate students presents systematic study of the topological structure of smooth manifolds, starting with elements of theory and concluding with method of surgery. 1993 edition. 288pp. 5⅜ x 8½. 0-486-46244-7

VECTOR AND TENSOR ANALYSIS WITH APPLICATIONS, A. I. Borisenko and I. E. Tarapov. Concise introduction. Worked-out problems, solutions, exercises. 257pp. 5⅜ x 8¼. 0-486-63833-2

AN INTRODUCTION TO ORDINARY DIFFERENTIAL EQUATIONS, Earl A. Coddington. A thorough and systematic first course in elementary differential equations for undergraduates in mathematics and science, with many exercises and problems (with answers). Index. 304pp. 5⅜ x 8½. 0-486-65942-9

FOURIER SERIES AND ORTHOGONAL FUNCTIONS, Harry F. Davis. An incisive text combining theory and practical example to introduce Fourier series, orthogonal functions and applications of the Fourier method to boundary-value problems. 570 exercises. Answers and notes. 416pp. 5⅜ x 8½. 0-486-65973-9

COMPUTABILITY AND UNSOLVABILITY, Martin Davis. Classic graduate-level introduction to theory of computability, usually referred to as theory of recurrent functions. New preface and appendix. 288pp. 5⅜ x 8½. 0-486-61471-9

AN INTRODUCTION TO MATHEMATICAL ANALYSIS, Robert A. Rankin. Dealing chiefly with functions of a single real variable, this text by a distinguished educator introduces limits, continuity, differentiability, integration, convergence of infinite series, double series, and infinite products. 1963 edition. 624pp. 5⅜ x 8½. 0-486-46251-X

METHODS OF NUMERICAL INTEGRATION (SECOND EDITION), Philip J. Davis and Philip Rabinowitz. Requiring only a background in calculus, this text covers approximate integration over finite and infinite intervals, error analysis, approximate integration in two or more dimensions, and automatic integration. 1984 edition. 624pp. 5⅜ x 8½. 0-486-45339-1

INTRODUCTION TO LINEAR ALGEBRA AND DIFFERENTIAL EQUATIONS, John W. Dettman. Excellent text covers complex numbers, determinants, orthonormal bases, Laplace transforms, much more. Exercises with solutions. Undergraduate level. 416pp. 5⅜ x 8½. 0-486-65191-6

RIEMANN'S ZETA FUNCTION, H. M. Edwards. Superb, high-level study of landmark 1859 publication entitled "On the Number of Primes Less Than a Given Magnitude" traces developments in mathematical theory that it inspired. xiv+315pp. 5⅜ x 8½. 0-486-41740-9

CALCULUS OF VARIATIONS WITH APPLICATIONS, George M. Ewing. Applications-oriented introduction to variational theory develops insight and promotes understanding of specialized books, research papers. Suitable for advanced undergraduate/graduate students as primary, supplementary text. 352pp. 5⅜ x 8½.
0-486-64856-7

MATHEMATICIAN'S DELIGHT, W. W. Sawyer. "Recommended with confidence" by *The Times Literary Supplement,* this lively survey was written by a renowned teacher. It starts with arithmetic and algebra, gradually proceeding to trigonometry and calculus. 1943 edition. 240pp. 5⅜ x 8½. 0-486-46240-4

ADVANCED EUCLIDEAN GEOMETRY, Roger A. Johnson. This classic text explores the geometry of the triangle and the circle, concentrating on extensions of Euclidean theory, and examining in detail many relatively recent theorems. 1929 edition. 336pp. 5⅜ x 8½. 0-486-46237-4

COUNTEREXAMPLES IN ANALYSIS, Bernard R. Gelbaum and John M. H. Olmsted. These counterexamples deal mostly with the part of analysis known as "real variables." The first half covers the real number system, and the second half encompasses higher dimensions. 1962 edition. xxiv+198pp. 5⅜ x 8½. 0-486-42875-3

CATASTROPHE THEORY FOR SCIENTISTS AND ENGINEERS, Robert Gilmore. Advanced-level treatment describes mathematics of theory grounded in the work of Poincaré, R. Thom, other mathematicians. Also important applications to problems in mathematics, physics, chemistry and engineering. 1981 edition. References. 28 tables. 397 black-and-white illustrations. xvii + 666pp. 6⅛ x 9¼.
0-486-67539-4

COMPLEX VARIABLES: Second Edition, Robert B. Ash and W. P. Novinger. Suitable for advanced undergraduates and graduate students, this newly revised treatment covers Cauchy theorem and its applications, analytic functions, and the prime number theorem. Numerous problems and solutions. 2004 edition. 224pp. 6½ x 9¼. 0-486-46250-1

NUMERICAL METHODS FOR SCIENTISTS AND ENGINEERS, Richard Hamming. Classic text stresses frequency approach in coverage of algorithms, polynomial approximation, Fourier approximation, exponential approximation, other topics. Revised and enlarged 2nd edition. 721pp. 5⅜ x 8½. 0-486-65241-6

INTRODUCTION TO NUMERICAL ANALYSIS (2nd Edition), F. B. Hildebrand. Classic, fundamental treatment covers computation, approximation, interpolation, numerical differentiation and integration, other topics. 150 new problems. 669pp. 5⅜ x 8½. 0-486-65363-3

MARKOV PROCESSES AND POTENTIAL THEORY, Robert M. Blumental and Ronald K. Getoor. This graduate-level text explores the relationship between Markov processes and potential theory in terms of excessive functions, multiplicative functionals and subprocesses, additive functionals and their potentials, and dual processes. 1968 edition. 320pp. 5⅜ x 8½. 0-486-46263-3

ABSTRACT SETS AND FINITE ORDINALS: An Introduction to the Study of Set Theory, G. B. Keene. This text unites logical and philosophical aspects of set theory in a manner intelligible to mathematicians without training in formal logic and to logicians without a mathematical background. 1961 edition. 112pp. 5⅜ x 8½.
0-486-46249-8

INTRODUCTORY REAL ANALYSIS, A.N. Kolmogorov, S. V. Fomin. Translated by Richard A. Silverman. Self-contained, evenly paced introduction to real and functional analysis. Some 350 problems. 403pp. 5⅜ x 8½. 0-486-61226-0

APPLIED ANALYSIS, Cornelius Lanczos. Classic work on analysis and design of finite processes for approximating solution of analytical problems. Algebraic equations, matrices, harmonic analysis, quadrature methods, much more. 559pp. 5⅜ x 8½. 0-486-65656-X

AN INTRODUCTION TO ALGEBRAIC STRUCTURES, Joseph Landin. Superb self-contained text covers "abstract algebra": sets and numbers, theory of groups, theory of rings, much more. Numerous well-chosen examples, exercises. 247pp. 5⅜ x 8½. 0-486-65940-2

QUALITATIVE THEORY OF DIFFERENTIAL EQUATIONS, V. V. Nemytskii and V.V. Stepanov. Classic graduate-level text by two prominent Soviet mathematicians covers classical differential equations as well as topological dynamics and ergodic theory. Bibliographies. 523pp. 5⅜ x 8½. 0-486-65954-2

THEORY OF MATRICES, Sam Perlis. Outstanding text covering rank, nonsingularity and inverses in connection with the development of canonical matrices under the relation of equivalence, and without the intervention of determinants. Includes exercises. 237pp. 5⅜ x 8½. 0-486-66810-X

INTRODUCTION TO ANALYSIS, Maxwell Rosenlicht. Unusually clear, accessible coverage of set theory, real number system, metric spaces, continuous functions, Riemann integration, multiple integrals, more. Wide range of problems. Undergraduate level. Bibliography. 254pp. 5⅜ x 8½. 0-486-65038-3

MODERN NONLINEAR EQUATIONS, Thomas L. Saaty. Emphasizes practical solution of problems; covers seven types of equations. ". . . a welcome contribution to the existing literature. . . ."–*Math Reviews*. 490pp. 5⅜ x 8½. 0-486-64232-1

MATRICES AND LINEAR ALGEBRA, Hans Schneider and George Phillip Barker. Basic textbook covers theory of matrices and its applications to systems of linear equations and related topics such as determinants, eigenvalues and differential equations. Numerous exercises. 432pp. 5⅜ x 8½. 0-486-66014-1

LINEAR ALGEBRA, Georgi E. Shilov. Determinants, linear spaces, matrix algebras, similar topics. For advanced undergraduates, graduates. Silverman translation. 387pp. 5⅜ x 8½. 0-486-63518-X

MATHEMATICAL METHODS OF GAME AND ECONOMIC THEORY: Revised Edition, Jean-Pierre Aubin. This text begins with optimization theory and convex analysis, followed by topics in game theory and mathematical economics, and concluding with an introduction to nonlinear analysis and control theory. 1982 edition. 656pp. 6⅛ x 9¼. 0-486-46265-X

SET THEORY AND LOGIC, Robert R. Stoll. Lucid introduction to unified theory of mathematical concepts. Set theory and logic seen as tools for conceptual understanding of real number system. 496pp. 5⅜ x 8¼. 0-486-63829-4

TENSOR CALCULUS, J.L. Synge and A. Schild. Widely used introductory text covers spaces and tensors, basic operations in Riemannian space, non-Riemannian spaces, etc. 324pp. 5⅜ x 8¼. 0-486-63612-7

ORDINARY DIFFERENTIAL EQUATIONS, Morris Tenenbaum and Harry Pollard. Exhaustive survey of ordinary differential equations for undergraduates in mathematics, engineering, science. Thorough analysis of theorems. Diagrams. Bibliography. Index. 818pp. 5⅜ x 8½. 0-486-64940-7

INTEGRAL EQUATIONS, F. G. Tricomi. Authoritative, well-written treatment of extremely useful mathematical tool with wide applications. Volterra Equations, Fredholm Equations, much more. Advanced undergraduate to graduate level. Exercises. Bibliography. 238pp. 5⅜ x 8½. 0-486-64828-1

FOURIER SERIES, Georgi P. Tolstov. Translated by Richard A. Silverman. A valuable addition to the literature on the subject, moving clearly from subject to subject and theorem to theorem. 107 problems, answers. 336pp. 5⅜ x 8½. 0-486-63317-9

INTRODUCTION TO MATHEMATICAL THINKING, Friedrich Waismann. Examinations of arithmetic, geometry, and theory of integers; rational and natural numbers; complete induction; limit and point of accumulation; remarkable curves; complex and hypercomplex numbers, more. 1959 ed. 27 figures. xii+260pp. 5⅜ x 8½. 0-486-42804-8

THE RADON TRANSFORM AND SOME OF ITS APPLICATIONS, Stanley R. Deans. Of value to mathematicians, physicists, and engineers, this excellent introduction covers both theory and applications, including a rich array of examples and literature. Revised and updated by the author. 1993 edition. 304pp. 6⅛ x 9¼. 0-486-46241-2

CALCULUS OF VARIATIONS, Robert Weinstock. Basic introduction covering isoperimetric problems, theory of elasticity, quantum mechanics, electrostatics, etc. Exercises throughout. 326pp. 5⅜ x 8½. 0-486-63069-2

THE CONTINUUM: A CRITICAL EXAMINATION OF THE FOUNDATION OF ANALYSIS, Hermann Weyl. Classic of 20th-century foundational research deals with the conceptual problem posed by the continuum. 156pp. 5⅜ x 8½. 0-486-67982-9

CHALLENGING MATHEMATICAL PROBLEMS WITH ELEMENTARY SOLUTIONS, A. M. Yaglom and I. M. Yaglom. Over 170 challenging problems on probability theory, combinatorial analysis, points and lines, topology, convex polygons, many other topics. Solutions. Total of 445pp. 5⅜ x 8½. Two-vol. set.
Vol. I: 0-486-65536-9 Vol. II: 0-486-65537-7

INTRODUCTION TO PARTIAL DIFFERENTIAL EQUATIONS WITH APPLICATIONS, E. C. Zachmanoglou and Dale W. Thoe. Essentials of partial differential equations applied to common problems in engineering and the physical sciences. Problems and answers. 416pp. 5⅜ x 8½. 0-486-65251-3

STOCHASTIC PROCESSES AND FILTERING THEORY, Andrew H. Jazwinski. This unified treatment presents material previously available only in journals, and in terms accessible to engineering students. Although theory is emphasized, it discusses numerous practical applications as well. 1970 edition. 400pp. 5⅜ x 8½. 0-486-46274-9

Math–Decision Theory, Statistics, Probability

INTRODUCTION TO PROBABILITY, John E. Freund. Featured topics include permutations and factorials, probabilities and odds, frequency interpretation, mathematical expectation, decision-making, postulates of probability, rule of elimination, much more. Exercises with some solutions. Summary. 1973 edition. 247pp. 5⅜ x 8½. 0-486-67549-1

STATISTICAL AND INDUCTIVE PROBABILITIES, Hugues Leblanc. This treatment addresses a decades-old dispute among probability theorists, asserting that both statistical and inductive probabilities may be treated as sentence-theoretic measurements, and that the latter qualify as estimates of the former. 1962 edition. 160pp. 5⅜ x 8½. 0-486-44980-7

APPLIED MULTIVARIATE ANALYSIS: Using Bayesian and Frequentist Methods of Inference, Second Edition, S. James Press. This two-part treatment deals with foundations as well as models and applications. Topics include continuous multivariate distributions; regression and analysis of variance; factor analysis and latent structure analysis; and structuring multivariate populations. 1982 edition. 692pp. 5⅜ x 8½. 0-486-44236-5

LINEAR PROGRAMMING AND ECONOMIC ANALYSIS, Robert Dorfman, Paul A. Samuelson and Robert M. Solow. First comprehensive treatment of linear programming in standard economic analysis. Game theory, modern welfare economics, Leontief input-output, more. 525pp. 5⅜ x 8½. 0-486-65491-5

PROBABILITY: AN INTRODUCTION, Samuel Goldberg. Excellent basic text covers set theory, probability theory for finite sample spaces, binomial theorem, much more. 360 problems. Bibliographies. 322pp. 5⅜ x 8½. 0-486-65252-1

GAMES AND DECISIONS: INTRODUCTION AND CRITICAL SURVEY, R. Duncan Luce and Howard Raiffa. Superb nontechnical introduction to game theory, primarily applied to social sciences. Utility theory, zero-sum games, n-person games, decision-making, much more. Bibliography. 509pp. 5⅜ x 8½. 0-486-65943-7

INTRODUCTION TO THE THEORY OF GAMES, J. C. C. McKinsey. This comprehensive overview of the mathematical theory of games illustrates applications to situations involving conflicts of interest, including economic, social, political, and military contexts. Appropriate for advanced undergraduate and graduate courses; advanced calculus a prerequisite. 1952 ed. x+372pp. 5⅜ x 8½. 0-486-42811-7

FIFTY CHALLENGING PROBLEMS IN PROBABILITY WITH SOLUTIONS, Frederick Mosteller. Remarkable puzzlers, graded in difficulty, illustrate elementary and advanced aspects of probability. Detailed solutions. 88pp. 5⅜ x 8½. 0-486-65355-2

PROBABILITY THEORY: A CONCISE COURSE, Y. A. Rozanov. Highly readable, self-contained introduction covers combination of events, dependent events, Bernoulli trials, etc. 148pp. 5⅝ x 8¼. 0-486-63544-9

THE STATISTICAL ANALYSIS OF EXPERIMENTAL DATA, John Mandel. First half of book presents fundamental mathematical definitions, concepts and facts while remaining half deals with statistics primarily as an interpretive tool. Well-written text, numerous worked examples with step-by-step presentation. Includes 116 tables. 448pp. 5⅜ x 8½. 0-486-64666-1

Math–Geometry and Topology

ELEMENTARY CONCEPTS OF TOPOLOGY, Paul Alexandroff. Elegant, intuitive approach to topology from set-theoretic topology to Betti groups; how concepts of topology are useful in math and physics. 25 figures. 57pp. 5⅜ x 8½. 0-486-60747-X

A LONG WAY FROM EUCLID, Constance Reid. Lively guide by a prominent historian focuses on the role of Euclid's Elements in subsequent mathematical developments. Elementary algebra and plane geometry are sole prerequisites. 80 drawings. 1963 edition. 304pp. 5⅜ x 8½. 0-486-43613-6

EXPERIMENTS IN TOPOLOGY, Stephen Barr. Classic, lively explanation of one of the byways of mathematics. Klein bottles, Moebius strips, projective planes, map coloring, problem of the Koenigsberg bridges, much more, described with clarity and wit. 43 figures. 210pp. 5⅜ x 8½. 0-486-25933-1

THE GEOMETRY OF RENÉ DESCARTES, René Descartes. The great work founded analytical geometry. Original French text, Descartes's own diagrams, together with definitive Smith-Latham translation. 244pp. 5⅜ x 8½. 0-486-60068-8

EUCLIDEAN GEOMETRY AND TRANSFORMATIONS, Clayton W. Dodge. This introduction to Euclidean geometry emphasizes transformations, particularly isometries and similarities. Suitable for undergraduate courses, it includes numerous examples, many with detailed answers. 1972 ed. viii+296pp. 6⅛ x 9¼. 0-486-43476-1

EXCURSIONS IN GEOMETRY, C. Stanley Ogilvy. A straightedge, compass, and a little thought are all that's needed to discover the intellectual excitement of geometry. Harmonic division and Apollonian circles, inversive geometry, hexlet, Golden Section, more. 132 illustrations. 192pp. 5⅜ x 8½. 0-486-26530-7

THE THIRTEEN BOOKS OF EUCLID'S ELEMENTS, translated with introduction and commentary by Sir Thomas L. Heath. Definitive edition. Textual and linguistic notes, mathematical analysis. 2,500 years of critical commentary. Unabridged. 1,414pp. 5⅜ x 8½. Three-vol. set.
Vol. I: 0-486-60088-2 Vol. II: 0-486-60089-0 Vol. III: 0-486-60090-4

SPACE AND GEOMETRY: IN THE LIGHT OF PHYSIOLOGICAL, PSYCHOLOGICAL AND PHYSICAL INQUIRY, Ernst Mach. Three essays by an eminent philosopher and scientist explore the nature, origin, and development of our concepts of space, with a distinctness and precision suitable for undergraduate students and other readers. 1906 ed. vi+148pp. 5⅜ x 8½. 0-486-43909-7

GEOMETRY OF COMPLEX NUMBERS, Hans Schwerdtfeger. Illuminating, widely praised book on analytic geometry of circles, the Moebius transformation, and two-dimensional non-Euclidean geometries. 200pp. 5⅝ x 8¼. 0-486-63830-8

DIFFERENTIAL GEOMETRY, Heinrich W. Guggenheimer. Local differential geometry as an application of advanced calculus and linear algebra. Curvature, transformation groups, surfaces, more. Exercises. 62 figures. 378pp. 5⅜ x 8½. 0-486-63433-7

History of Math

THE WORKS OF ARCHIMEDES, Archimedes (T. L. Heath, ed.). Topics include the famous problems of the ratio of the areas of a cylinder and an inscribed sphere; the measurement of a circle; the properties of conoids, spheroids, and spirals; and the quadrature of the parabola. Informative introduction. clxxxvi+326pp. 5⅜ x 8½. 0-486-42084-1

A SHORT ACCOUNT OF THE HISTORY OF MATHEMATICS, W. W. Rouse Ball. One of clearest, most authoritative surveys from the Egyptians and Phoenicians through 19th-century figures such as Grassman, Galois, Riemann. Fourth edition. 522pp. 5⅜ x 8½. 0-486-20630-0

THE HISTORY OF THE CALCULUS AND ITS CONCEPTUAL DEVELOPMENT, Carl B. Boyer. Origins in antiquity, medieval contributions, work of Newton, Leibniz, rigorous formulation. Treatment is verbal. 346pp. 5⅜ x 8½. 0-486-60509-4

THE HISTORICAL ROOTS OF ELEMENTARY MATHEMATICS, Lucas N. H. Bunt, Phillip S. Jones, and Jack D. Bedient. Fundamental underpinnings of modern arithmetic, algebra, geometry and number systems derived from ancient civilizations. 320pp. 5⅜ x 8½. 0-486-25563-8

THE HISTORY OF THE CALCULUS AND ITS CONCEPTUAL DEVELOPMENT, Carl B. Boyer. Fluent description of the development of both the integral and differential calculus–its early beginnings in antiquity, medieval contributions, and a consideration of Newton and Leibniz. 368pp. 5⅜ x 8½. 0-486-60509-4

GAMES, GODS & GAMBLING: A HISTORY OF PROBABILITY AND STATISTICAL IDEAS, F. N. David. Episodes from the lives of Galileo, Fermat, Pascal, and others illustrate this fascinating account of the roots of mathematics. Features thought-provoking references to classics, archaeology, biography, poetry. 1962 edition. 304pp. 5⅜ x 8½. (Available in U.S. only.) 0-486-40023-9

OF MEN AND NUMBERS: THE STORY OF THE GREAT MATHEMATICIANS, Jane Muir. Fascinating accounts of the lives and accomplishments of history's greatest mathematical minds–Pythagoras, Descartes, Euler, Pascal, Cantor, many more. Anecdotal, illuminating. 30 diagrams. Bibliography. 256pp. 5⅜ x 8½. 0-486-28973-7

HISTORY OF MATHEMATICS, David E. Smith. Nontechnical survey from ancient Greece and Orient to late 19th century; evolution of arithmetic, geometry, trigonometry, calculating devices, algebra, the calculus. 362 illustrations. 1,355pp. 5⅜ x 8½. Two-vol. set. Vol. I: 0-486-20429-4 Vol. II: 0-486-20430-8

A CONCISE HISTORY OF MATHEMATICS, Dirk J. Struik. The best brief history of mathematics. Stresses origins and covers every major figure from ancient Near East to 19th century. 41 illustrations. 195pp. 5⅜ x 8½. 0-486-60255-9

Physics

OPTICAL RESONANCE AND TWO-LEVEL ATOMS, L. Allen and J. H. Eberly. Clear, comprehensive introduction to basic principles behind all quantum optical resonance phenomena. 53 illustrations. Preface. Index. 256pp. 5⅜ x 8½. 0-486-65533-4

QUANTUM THEORY, David Bohm. This advanced undergraduate-level text presents the quantum theory in terms of qualitative and imaginative concepts, followed by specific applications worked out in mathematical detail. Preface. Index. 655pp. 5⅜ x 8½. 0-486-65969-0

ATOMIC PHYSICS (8th EDITION), Max Born. Nobel laureate's lucid treatment of kinetic theory of gases, elementary particles, nuclear atom, wave-corpuscles, atomic structure and spectral lines, much more. Over 40 appendices, bibliography. 495pp. 5⅜ x 8½. 0-486-65984-4

A SOPHISTICATE'S PRIMER OF RELATIVITY, P. W. Bridgman. Geared toward readers already acquainted with special relativity, this book transcends the view of theory as a working tool to answer natural questions: What is a frame of reference? What is a "law of nature"? What is the role of the "observer"? Extensive treatment, written in terms accessible to those without a scientific background. 1983 ed. xlviii+172pp. 5⅜ x 8½. 0-486-42549-5

AN INTRODUCTION TO HAMILTONIAN OPTICS, H. A. Buchdahl. Detailed account of the Hamiltonian treatment of aberration theory in geometrical optics. Many classes of optical systems defined in terms of the symmetries they possess. Problems with detailed solutions. 1970 edition. xv + 360pp. 5⅜ x 8½. 0-486-67597-1

PRIMER OF QUANTUM MECHANICS, Marvin Chester. Introductory text examines the classical quantum bead on a track: its state and representations; operator eigenvalues; harmonic oscillator and bound bead in a symmetric force field; and bead in a spherical shell. Other topics include spin, matrices, and the structure of quantum mechanics; the simplest atom; indistinguishable particles; and stationary-state perturbation theory. 1992 ed. xiv+314pp. 6⅛ x 9¼. 0-486-42878-8

LECTURES ON QUANTUM MECHANICS, Paul A. M. Dirac. Four concise, brilliant lectures on mathematical methods in quantum mechanics from Nobel Prize-winning quantum pioneer build on idea of visualizing quantum theory through the use of classical mechanics. 96pp. 5⅜ x 8½. 0-486-41713-1

THIRTY YEARS THAT SHOOK PHYSICS: THE STORY OF QUANTUM THEORY, George Gamow. Lucid, accessible introduction to influential theory of energy and matter. Careful explanations of Dirac's anti-particles, Bohr's model of the atom, much more. 12 plates. Numerous drawings. 240pp. 5⅜ x 8½. 0-486-24895-X

ELECTRONIC STRUCTURE AND THE PROPERTIES OF SOLIDS: THE PHYSICS OF THE CHEMICAL BOND, Walter A. Harrison. Innovative text offers basic understanding of the electronic structure of covalent and ionic solids, simple metals, transition metals and their compounds. Problems. 1980 edition. 582pp. 6⅛ x 9¼. 0-486-66021-4

HYDRODYNAMIC AND HYDROMAGNETIC STABILITY, S. Chandrasekhar. Lucid examination of the Rayleigh-Benard problem; clear coverage of the theory of instabilities causing convection. 704pp. 5⅝ x 8¼. 0-486-64071-X

INVESTIGATIONS ON THE THEORY OF THE BROWNIAN MOVEMENT, Albert Einstein. Five papers (1905–8) investigating dynamics of Brownian motion and evolving elementary theory. Notes by R. Fürth. 122pp. 5⅜ x 8½. 0-486-60304-0

THE PHYSICS OF WAVES, William C. Elmore and Mark A. Heald. Unique overview of classical wave theory. Acoustics, optics, electromagnetic radiation, more. Ideal as classroom text or for self-study. Problems. 477pp. 5⅜ x 8½. 0-486-64926-1

GRAVITY, George Gamow. Distinguished physicist and teacher takes reader-friendly look at three scientists whose work unlocked many of the mysteries behind the laws of physics: Galileo, Newton, and Einstein. Most of the book focuses on Newton's ideas, with a concluding chapter on post-Einsteinian speculations concerning the relationship between gravity and other physical phenomena. 160pp. 5⅜ x 8½. 0-486-42563-0

PHYSICAL PRINCIPLES OF THE QUANTUM THEORY, Werner Heisenberg. Nobel Laureate discusses quantum theory, uncertainty, wave mechanics, work of Dirac, Schroedinger, Compton, Wilson, Einstein, etc. 184pp. 5⅜ x 8½. 0-486-60113-7

ATOMIC SPECTRA AND ATOMIC STRUCTURE, Gerhard Herzberg. One of best introductions; especially for specialist in other fields. Treatment is physical rather than mathematical. 80 illustrations. 257pp. 5⅜ x 8½. 0-486-60115-3

AN INTRODUCTION TO STATISTICAL THERMODYNAMICS, Terrell L. Hill. Excellent basic text offers wide-ranging coverage of quantum statistical mechanics, systems of interacting molecules, quantum statistics, more. 523pp. 5⅜ x 8½. 0-486-65242-4

THEORETICAL PHYSICS, Georg Joos, with Ira M. Freeman. Classic overview covers essential math, mechanics, electromagnetic theory, thermodynamics, quantum mechanics, nuclear physics, other topics. First paperback edition. xxiii + 885pp. 5⅜ x 8½. 0-486-65227-0

PROBLEMS AND SOLUTIONS IN QUANTUM CHEMISTRY AND PHYSICS, Charles S. Johnson, Jr. and Lee G. Pedersen. Unusually varied problems, detailed solutions in coverage of quantum mechanics, wave mechanics, angular momentum, molecular spectroscopy, more. 280 problems plus 139 supplementary exercises. 430pp. 6½ x 9¼. 0-486-65236-X

THEORETICAL SOLID STATE PHYSICS, Vol. 1: Perfect Lattices in Equilibrium; Vol. II: Non-Equilibrium and Disorder, William Jones and Norman H. March. Monumental reference work covers fundamental theory of equilibrium properties of perfect crystalline solids, non-equilibrium properties, defects and disordered systems. Appendices. Problems. Preface. Diagrams. Index. Bibliography. Total of 1,301pp. 5⅜ x 8½. Two volumes. Vol. I: 0-486-65015-4 Vol. II: 0-486-65016-2

WHAT IS RELATIVITY? L. D. Landau and G. B. Rumer. Written by a Nobel Prize physicist and his distinguished colleague, this compelling book explains the special theory of relativity to readers with no scientific background, using such familiar objects as trains, rulers, and clocks. 1960 ed. vi+72pp. 5⅜ x 8½. 0-486-42806-0

CPSIA information can be obtained
at www.ICGtesting.com
Printed in the USA
FSHW01n0604280918
52601FS